Lecture Notes in Computer Science 11960

More information about this series at http://www.springer.com/series/8183

Marina L. Gavrilova · C. J. Kenneth Tan ·
Khalid Saeed · Nabendu Chaki (Eds.)

Transactions on Computational Science XXXV

Special Issue on Signal Processing and Security
in Distributed Systems

 Springer

Editors-in-Chief
Marina L. Gavrilova
University of Calgary
Calgary, AB, Canada

C. J. Kenneth Tan
Sardina Systems OÜ
Tallinn, Estonia

Guest Editors
Khalid Saeed
Bialystok University of Technology
Bialystok, Poland

Nabendu Chaki
University of Calcutta
Kolkata, India

ISSN 0302-9743　　　　　　　ISSN 1611-3349　(electronic)
Lecture Notes in Computer Science
ISSN 1866-4733　　　　　　　ISSN 1866-4741　(electronic)
Transactions on Computational Science
ISBN 978-3-662-61091-6　　　ISBN 978-3-662-61092-3　(eBook)
https://doi.org/10.1007/978-3-662-61092-3

This Springer imprint is published by the registered company Springer-Verlag GmbH, DE
part of Springer Nature
The registered company address is: Heidelberger Platz 3, 14197 Berlin, Germany

LNCS Transactions on Computational Science

Computational science, an emerging and increasingly vital field, is now widely recognized as an integral part of scientific and technical investigations, affecting researchers and practitioners in areas ranging from aerospace and automotive research to biochemistry, electronics, geosciences, mathematics, and physics. Computer systems research and the exploitation of applied research naturally complement each other. The increased complexity of many challenges in computational science demands the use of supercomputing, parallel processing, sophisticated algorithms, and advanced system software and architecture. It is therefore invaluable to have input by systems research experts in applied computational science research.

Transactions on Computational Science focuses on original high-quality research in the realm of computational science in parallel and distributed environments, also encompassing the underlying theoretical foundations and the applications of large-scale computation.

The journal offers practitioners and researchers the opportunity to share computational techniques and solutions in this area, to identify new issues, and to shape future directions for research, and it enables industrial users to apply leading-edge, large-scale, high-performance computational methods.

In addition to addressing various research and application issues, the journal aims to present material that is validated – crucial to the application and advancement of the research conducted in academic and industrial settings. In this spirit, the journal focuses on publications that present results and computational techniques that are verifiable.

Scope

The scope of the journal includes, but is not limited to, the following computational methods and applications:

- Aeronautics and Aerospace
- Astrophysics
- Big Data Analytics
- Bioinformatics
- Biometric Technologies
- Climate and Weather Modeling
- Communication and Data Networks
- Compilers and Operating Systems
- Computer Graphics
- Computational Biology
- Computational Chemistry
- Computational Finance and Econometrics
- Computational Fluid Dynamics

- Computational Geometry
- Computational Number Theory
- Data Representation and Storage
- Data Mining and Data Warehousing
- Information and Online Security
- Grid Computing
- Hardware/Software Co-design
- High-Performance Computing
- Image and Video Processing
- Information Systems
- Information Retrieval
- Modeling and Simulations
- Mobile Computing
- Numerical and Scientific Computing
- Parallel and Distributed Computing
- Robotics and Navigation
- Supercomputing
- System-on-Chip Design and Engineering
- Virtual Reality and Cyberworlds
- Visualization

Editorial

The *Transactions on Computational Science* journal is published as part of the Springer series *Lecture Notes in Computer Science*, and is devoted to the gamut of computational science issues, from theoretical aspects to application-dependent studies and the validation of emerging technologies.

The journal focuses on original high-quality research in the realm of computational science in parallel and distributed environments, encompassing the facilitating theoretical foundations and the applications of large-scale computations and massive data processing. Practitioners and researchers share computational techniques and solutions in the area, identify new issues, and shape future directions for research, as well as enable industrial users to apply the presented techniques.

The current volume is devoted to the area of Signal Processing and Security in Distributed Systems and is edited by Khalid Saeed and Nabendu Chaki. This special issue contains seven selected papers, including a position paper and the selected papers invited following the 5th and 6th Doctoral Symposium on Applied Computation and Security Systems (ACSS). All of the accepted papers have been peer-reviewed.

We would like to extend our sincere appreciation to the special issue guest editors, Khalid Saeed and Nabendu Chaki, for their dedication and insights in preparing this high-quality special issue. We would also like to thank all of the authors for submitting their papers to the special issue and the associate editors and referees for their valuable work.

We do hope that the fine collection of papers presented in this special issue will be a valuable resource for *Transactions on Computational Science* readers and will stimulate further research in the vibrant area of computational science applications.

November 2019

Marina L. Gavrilova
C. J. Kenneth Tan

Guest Editors' Preface

The current volume is devoted to the area of Signal Processing and Security in Distributed Systems and is edited by Khalid Saeed and Nabendu Chaki. This special issue contains seven selected papers, three of which are extended from selected works presented at the 5th Doctoral Symposium on Applied Computation and Security Systems (ACSS) held in 2018, and three of the papers are extended versions of works presented at the 6th ACSS held in 2019. The other contribution is a position paper, invited by the guest editors, from Prof. Nobuyuki Nishiuchi of Tokyo Metropolitan University Hino, Tokyo, Japan. The authors for the selected papers are from different countries in Europe and Asia.

ACSS aims to facilitate PhD scholars enrolled with universities and research institutes around the world to present and discuss part of their research work with peers in their fields. Each contributed paper presented at ACSS must have at least one enrolled PhD student as the first author of the submission and his/her supervisor(s) as co-author(s). ACSS has been co-organized annually since 2014 by the University of Calcutta, India, together with Ca' Foscari University, Italy, and Bialystok University of Technology, Poland. As indicated in the name of symposium, security remains one of the primary focus areas for this annual meeting.

A two-volume post-symposium book has been published by Springer in the AISC series since the inception of ACSS in 2014. The papers selected for this special issue indicate the excellence that marks the success of ACSS in bringing the PhD scholars into this forum for the exchange of ideas toward achieving greater scientific goals.

The first paper in this special issue is "Classification of Visual Attention Level During Target Gazing Using Microsaccades" authored by Soichiro Yokoo, Nobuyuki Nishiuchi, and Kimihiro Yamanaka. The authors proposed a novel system to classify visual attention levels during visual target gazing using microsaccades. The authors presented a proof of the concept by an experimental study involving ten subjects performing three tasks requiring different levels of visual attention. Eventually, this would help to assess the degree of visual attention paid by the subject while gazing at the target.

In the second paper titled "Multiscale Analysis of Textual Content Using Eyegaze" by Aniruddha Sinha, Rikayan Chaki, Bikram Kumar De, Rajlakshmi Guha, Sanjoy Kumar Saha, and Anupam Basu, the authors presented an approach for analyzing textual content in various scales using eyegaze. The scales included individual fixation characteristics, saccades and fixations within a line, and an overall difficulty score of the content.

The third paper in this special issue is titled "In-Car eCall Device for Automatic Accident Detection, Passengers Counting and Alarming." This is authored by Anna Lupinska-Dubicka, Marek Tabedzki, Marcin Adamski, Mariusz Rybnik, Mirosław Omieljanowicz, Maciej Szymkowski, Marek Gruszewski, Adam Klimowicz, Grzegorz Rubin, and Khalid Saeed. In this work, the authors have presented the details of a concept for a special eCall device installed in vehicles. These eCall devices are part of the European eSafety initiative and aim to improve the safety and efficiency of road transport. The proposed system will eventually be able to detect a road accident, indicate the number of vehicle's occupants, and send this information to a pre-fixed emergency service provider.

The fourth work is titled "Volumetric Density of Triangulated Range Images for Face Recognition" and is authored by Koushik Dutta, Debotosh Bhattacharjee, and Mita Nasipuri. The paper contributes towards developing a robust 3D face recognition system. A volumetric space has been created on some distinct triangular regions of the 3D range face image. The proposed 3D face recognition system has three major components. At first, seven significant landmarks are detected on the face. Secondly, the nose tip and two more individual landmarks are used to create a triangular region. Third, a plane is assumed at the nose tip level for representing the volumetric space. Three well-known 3D face databases: Frav3D, Bosphorus, and Gavabdb are used to validate the concept. On these databases, the system acquires very high recognition rates using kNN and SVM classifiers separately.

The fifth paper in this special issues, "Combining Merkle Hash Tree and Chaotic Cryptography for Secure Data Fusion in IoT" authored by Nashreen Nesa and Indrajit Banerjee, puts forward a security protocol that integrates authentication of a large number of devices and encryption of the generated data from these devices in an IoT environment. Maintaining the privacy of the users and access control on sensitive information was the primary focus of this study. The authors have improvised on the Merkle Hash Tree and utilized the concepts of Chaos theory towards proposing a new encryption algorithm.

The sixth paper in this volume is titled "A Deployment Framework for Ensuring Business Compliance Using Goal Models" and is authored by Novarun Deb, Mandira Roy, Surochita Pal, Ankita Bhaumik, and Nabendu Chaki. The work is in the field of goal model based requirements engineering by transforming a sequence agnostic goal model into a finite state model (FSM) and then checking them against temporal properties (in CTL). There are existing guidelines for generating such FSMs to provide a formal approach to prune a noncompliant FSM and generate FSMs that satisfy specified temporal properties. In this paper, the authors present a new version of a framework called i*ToNuSMV 3.0. The working of the framework is demonstrated with the help of some use cases.

The last article in this volume "A Methodology for Root-Causing In-field Attacks on Microfluidic Executions" by Pushpita Roy, Ansuman Banerjee, and Bhargab Bhattacharya presents a novel scheme for root-causing assay manipulation attacks for actuations on digital microfluidic biochips that manifest as errors after execution. The authors have used a functionally correct reaction sequence graph as the reference model from which the actuation sequence, to be implemented on a target Lab-on-chip, is synthesized. Program slicing is used to debug and locate the root cause of errors. Experimental observations are given, establishing the effectiveness of the proposed method in identifying the errors. The work may turn out to be pioneering research in improving the security and trustworthiness of microfluidic biochips.

We take this opportunity to express our heartfelt thanks and indebtedness to Prof. Marina Gavrilova, the Editor-in-Chief of *Transactions on Computational Science*, for her continual guidance in making this special issue. We thank the journal's editorial staff for their support and hard work towards developing the volume. We are grateful to the authors for their high-quality contributions and cooperation in preparing this special issue. With a deep sense of gratitude, we appreciate the support from the ACSS Program Committee members and reviewers, especially from Prof. Agostino Cortesi and Prof. Rituparna Chaki for choosing the best of the submissions from ACSS 2018 and ACSS 2019. They did an excellent job in encouraging authors to make significant extensions and advancements of the original works that were presented during the symposium. Our special thanks to all the organizers and sponsors of ACSS 2019 for their support in organizing the symposium. The culmination of their efforts at the grass-root level has led to this special issue.

This foreword will remain incomplete without a mention to the readers of this special issue of *Transactions on Computational Science* journal. Many thanks to all of them!

November 2019   Khalid Saeed
                                                                     Nabendu Chaki

Contents

Classification of Visual Attention Level During Target Gazing Using Microsaccades

Soichiro Yokoo[1](✉), Nobuyuki Nishiuchi[1], and Kimihiro Yamanaka[2]

[1] Graduate School of Systems Design, Faculty of Computer Science,
Tokyo Metropolitan University, 6-6 Asahigaoka, Hino, Tokyo 191-0065, Japan
yokoo-soichiro@ed.tmu.ac.jp, nnishiuchi@tmu.ac.jp
[2] Graduate School of Natural Science, Faculty of Intelligence and Informatics,
Konan University, 8-9-1 Okamoto, Higashi-Nada, Kobe, Hyogo 658-8501, Japan
kiyamana@konan-u.ac.jp

Abstract. In the previous researches on microsaccades, a typical and basic experimental method involves the following: a dot is displayed as a visual target in the middle or near the middle of a monitor, and eye movements of the subject are measured, and then the number of microsaccades is extracted from the measured eye movements. However, it is difficult to determine whether or not the subject is paying visual attention while gazing at the target, and the degree of visual attention paid by the subject. This paper proposes a system that uses microsaccades to classify visual attention levels during visual target gazing. In our experiment, ten subjects performed three tasks requiring different levels of visual attention. Microsaccades were measured and the number of microsaccades was extracted for each task. Statistical analysis showed that the number of microsaccades differed among the tasks. Our results suggest that visual attention level can be classified by the number of microsaccades.

Keywords: Microsaccade · Visual attention · Eye tracking

1 Introduction

The development of sensing technology has improved the ease and specificity of measurement of eye movements. Eye movements are a focus of many areas, including education, sport, human interfaces, and safety technology related to driving. Therefore, quantitative measurement and evaluation of eye movements is important for various research fields.

Eye movements can be classified into 2 types: conscious and unconscious eye movements [1]. Saccades are a type of conscious eye movement that are thought to play an important role in visual attention or interest because they facilitate the projection of a visual subject onto the central fovea. That is, saccades enable the visual behavior of shifting attention. In contrast, fixational eye movements are a type of unconscious eye movement where small involuntary eye movements are performed during visual fixation. Fixational eye movements can be categorized as microsaccades, drift or tremor. While microsaccades were initially thought to only impart change for the purpose of stimulating visual cells to keep scenes stationary, reports have shown

© Springer-Verlag GmbH Germany, part of Springer Nature 2020
M. L. Gavrilova et al. (Eds.): Trans. on Comput. Sci. XXXV, LNCS 11960, pp. 1–11, 2020.
https://doi.org/10.1007/978-3-662-61092-3_1

that the frequency and direction of microsaccades is actually related to visual attention [3–10].

A typical and basic experimental method that has been widely used in research on visual attention using microsaccades involves the following: (1) a dot is displayed as a visual target in the middle or near the middle of a monitor, (2) eye movements of the subject are measured, and (3) the number of microsaccades is extracted from the measured eye movements [3–9]. The advantage of this method is that its simple visual target minimizes measurement noise. However, it is difficult to determine whether or not the subject is paying visual attention while gazing at the target, and the degree of visual attention paid by the subject.

Therefore, we attempted to classify subjects' visual attention using a new experimental approach involving the analysis of microsaccades data. Our findings suggest that the visual attention level can be classified according to the number of microsaccades.

This paper is organized as follows: Sect. 2 summarizes related research; Sect. 3 describes the methodology of our experiments and analysis; Sect. 4 discusses the results of the experiment; and Sect. 5 summarizes our conclusions.

2 Related Works

Studies have proposed that the brain active state can be quantitatively evaluated using fixational eye movements to examine brain function. For example, Kashiwara [11] reported that emotional stimulation that induces subjective pleasure and discomfort affects not only the area of the pupil, but also the appearance rate of microsaccades. Sakaino [12] showed that the structure and characteristics of the spatial frequency of the visual stimulus affect fixational eye movements. Moreover, Carpenter [13] found that microsaccades are controlled by the same neuronal mechanisms as those that control saccades, which are conscious eye movements.

Regarding the relationship between microsaccades and visual attention, a previous study reported that visual attention involved in concentrating gaze in the vicinity of a fixed target could be classified based on the frequency and amplitude of microsaccades [3]. Moreover, several studies have suggested that it is possible to classify the direction of attention using pre-stimulus cues [4, 8, 9]. Other studies have reported a relationship between the features of microsaccades and the shape of the visual target (dot, circle and cross) [6], and a relationship between attention level and its reaction latency [7]. Interestingly, a recent applied study demonstrated that experts and beginners of tennis can also be classified using microsaccades analysis [10]. Apart from this, however, few applied studies on microsaccades have been reported.

Although a close relationship between saccades and visual attention has been shown [14, 15], we are unaware of any specific analysis of the relationship between microsaccades and visual attention. This suggests that new experimental conditions may be needed to establish a quantitative evaluation method for visual attention using microsaccades.

A number of methods have been reported for the specific extraction of microsaccades from eye movements. Kohama [3, 16] reported that it is possible to detect microsaccades using a data smoothing method by applying a wavelet transform. Noguchi [17] reported detection of microsaccades using a proposed non-linear filter that combines non-linear pre-processing based on order statistics and a low-pass derivative filter [17]. A new detection approach has also been proposed using neural networks such as convolutional neural networks [18] and recursive neural networks [19]. Among these methods, the method proposed by Engbert and Kliegl [8] is widely used in many studies on microsaccades [8–10, 20]. Their method involves (1) calculating the velocity of eye movements, (2) standardizing the velocity data using its average and standard deviation and (3) defining standardized data that is over the threshold as microsaccades. The data analysis in our study is based on this method.

3 Methodology

3.1 Experimental Environment

An eye tracking device (Tobii, Tobii Pro Spectrum, sampling rate: 600 Hz) and a 23.8-inch display (EIZO, FlexScan EV2451) were used. Measured eye movement data were saved to a PC (Dell, Precision 7720, OS: Microsoft Windows 10, CPU: Intel Xeon) and analyzed using a data management application (Tobii, Eye Tracker Manager). A light-shielding wall was used during the measurement of eye movements to reduce visual disturbances, limiting visual stimulation to just the visual target. The organization of the equipment is shown in Fig. 1.

Ten subjects performed the tasks (7 males and 3 females; age 21–27 years). We explained the details of the experiment to the subjects before the measurement, and the subjects' eyesight in both eyes was confirmed to be over 20/30 (vision test) (Figs. 2 and 3).

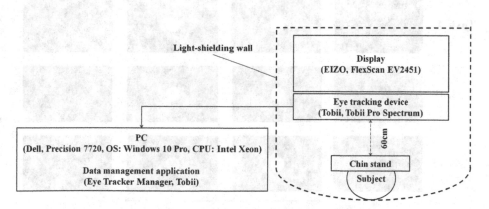

Fig. 1. Arrangement of measuring devices

Fig. 2. Eye tracking device (Tobii Pro Spectrum)

Fig. 3. Image of the experimental set up

The subjects were instructed to place their chin on a head-fixing device that was positioned 60 cm from the measurement device, and the height was adjusted so that the subjects' gaze fell naturally on the center of the display. The subject's body was not restrained using a belt or any equipment during the measurement.

3.2 Experimental Tasks

The experimental tasks are shown in Fig. 4. The subjects performed three tasks (Task A, B and C) in the experiment. Calibration was performed before each task. To eliminate sequential effects, the sequence of the three tasks was different for each subject. Further, the subjects were instructed not to blink while gazing at the target.

In Task A, subjects gazed at a white dot (on a black background) at a 1° viewing angle for 5 s. This task was assumed to require a low level of visual attention.

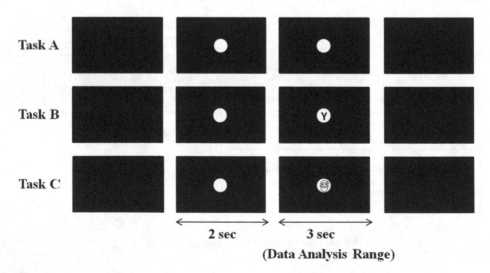

Fig. 4. Time sequence of visual target in each task

In Task B, the subjects gazed at the same white dot for 2 s before being shown a letter (Y) that was written over the white dot for 3 s. This task was assumed to require a middle level of visual attention. In Task C, the subjects gazed at the white dot for 2 s before being shown a line drawing (Santa Claus) that was written over the white dot for 3 s. This task was assumed to require a high level of visual attention.

Before and after each task, the display was dark. After Task B and C, the subjects were asked to identify the white dot, letter and line drawing that was written over the white dot. Each task was conducted 10 times for each subject. Each subject performed 30 trials in total. A 1° viewing angle of the visual target was used because the target is projected onto the central fovea at this angle [12].

3.3 Data Analysis

Engbert and Kliegl [8] defined microsaccades as eye movements with a velocity greater than the velocity threshold. We adopted this definition to extract microsaccades. The specific algorithm used was as follows:

(1) The time series of eye positions was transformed to velocity.

$$V_{xt} = \frac{X_{t+2} + X_{t+1} - X_{t-1} - X_{t-2}}{6\Delta t}, \quad V_{yt} = \frac{y_{t+2} + y_{t+1} - y_{t-1} - y_{t-2}}{6\Delta t}$$

$$(1)$$

(2) The scalar of the velocity was calculated.

$$V_{xyt} = \sqrt{V_{xt}^2 + V_{yt}^2}$$

$$(2)$$

(3) V_{xyt} was standardized using its standard deviation σ in the trials.
(4) The range was extracted if V_{xyt} was over the threshold $V_{th} = 3\sigma$.
(5) The range was detected as a microsaccade if it was continually over 6 ms (four samples in the current study satisfied this criteria).

For all eye movement acquisition data, data analysis was conducted using the above algorithm on only the final three seconds, after the initial 2 s had passed since presentation of the white dot target (Fig. 4). If error data (no detection of eye movements) were included, the percentage of errors was calculated. If the percentage of errors was greater than 5%, the data were not used and an additional trial was performed to make the total to 10 trials. If the percentage of errors was less than 5%, the error data were excluded.

The features of microsaccades and saccades are summarized in Table 1 [22]. The velocity and amplitude of saccades is greater than that of microsaccades. Therefore, inclusion of saccades in the analysis would prevent accurate extraction of microsaccades because the threshold would likely be higher. Therefore, eye movements with a velocity greater than 120°/sec were excluded based on Takahashi's method [10].

Table 1. Features of saccades and microsaccades [22]

	Microsaccade	Saccade
Velocity [deg/sec]	10–120	100–500
Amplitude [deg]	≤ 1	0.5–50
Duration time [msec]	10–30	20–80
Frequency [Hz]	0.33–3.33	2–3
Voluntary/Involuntary	Involuntary	Voluntary

4   Result

A scatter plot of the peak velocity (vertical axis) versus the amplitude (horizontal axis) of the extracted microsaccades is shown in Figs. 5, 6 and 7 for Task A, B and C, respectively. In addition, all subjects correctly identified the white dot, letter or line drawing.

Zuber et al. [23] reported that microsaccades exhibit linearity and a correlation between peak velocity and amplitude. This property has been used as a criterion in many studies to confirm the validity of the extracted microsaccades. While Zuber et al. did not describe their quantitative criteria, Ueda et al. [24] suggested that a quantitative criterion was that the correlation coefficient R^2 must be larger than 4/(the number of data points + 2). We used Ueda's criterion to confirm the validity of our extracted microsaccades. The extracted microsaccades for each task satisfied the criterion as shown in Figs. 5, 6 and 7.

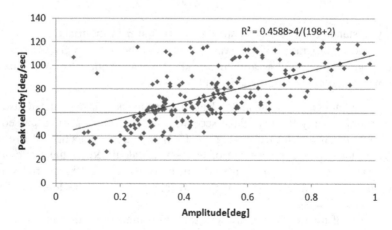

Fig. 5. Scatter plot of the peak velocity and amplitude of the extracted microsaccades. One hundred and ninety-eight microsaccades were extracted from the 10 subjects in Task A.

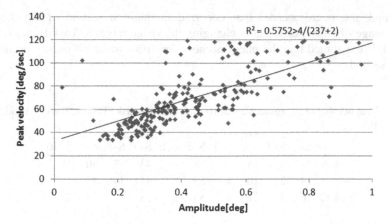

Fig. 6. Scatter plot of the peak velocity and amplitude of the extracted microsaccades. Two hundred and thirty-seven microsaccades were extracted from the 10 subjects in Task B.

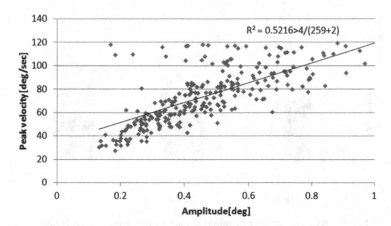

Fig. 7. Scatter plot of the peak velocity and amplitude of the extracted microsaccades. Two hundred and fifty-nine microsaccades were extracted from the 10 subjects in Task C.

According to Table 1, microsaccades and saccades are intermixed in the range 100–120°/sec. In Figs. 5, 6 and 7, a large number of data points did not show linearity or a correlation in the range 100–120°/sec. Therefore, subsequent analysis was performed after excluding data points in the range 100–120°/sec.

The total number of microsaccades in 10 trials (with peak velocities within the range 0–100°/sec) is shown in Table 2 for each task and subject. The average number of microsaccades in 10 trials is shown in Fig. 8 for each task.

One-way analysis of variance using the data in Table 2 showed no significant differences among the three tasks.

However, it is necessary to note that microsaccades also occur while viewing stationary scenes [2]. That is, there is a possibility that "microsaccades for viewing stationary scenes" and "microsaccades for visual attention" are mixed. On this background, the microsaccades that occurred in the first second of the 3-second analysis range (the first second after changing the visual target in Task B and C) were extracted. We assumed that the microsaccades for visual attention mostly occurred in the first second.

Table 2. Total number of microsaccades in 10 trials for each task (Task A, B and C) and subject (S–1 to S–10).

	S–1	S–2	S–3	S–4	S–5	S–6	S–7	S–8	S–9	S–10
Task A	14	3	16	4	38	6	43	3	30	1
Task B	14	2	10	4	55	6	41	19	37	1
Task C	16	7	21	0	52	3	33	39	35	7

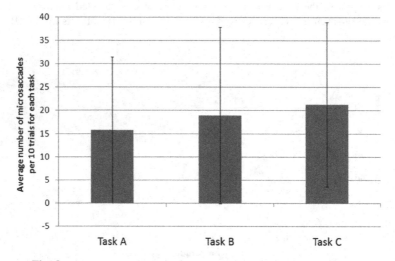

Fig. 8. Average number of microsaccades in 10 trials for each task.

The total number of microsaccades in the first second in 10 trials is shown in Table 3 for each task and subject. The average number of microsaccades in the first second in 10 trials is shown in Fig. 9 for each task. The results of one-way analysis of variance using the data in Table 3 are shown in Table 4. The results of the analysis approached significance ($p < 0.10$), suggesting a trend towards a relationship between the level of visual attention and the number of microsaccades. Multiple comparison showed that there was a significant difference in the number of microsaccades performed between Task A and Task C ($p < 0.05$, LSD).

One-way analysis of variance was also conducted using data obtained in the first 2 s of the 3-second analysis range; however, no significant differences were observed. That we only observed a significant difference using data from the first second may be because doing so excluded microsaccades performed for viewing stationary scenes and included only those used for visual attention. This is in contrast to the second and third seconds, when the subjects no longer had to recognize the visual target. Additionally, the lack of a significant difference between Task A and Task B may be due to the fact that the visual attention level was not linear across the tasks (Task A: white dot, Task B: white dot and letter, Task C: white dot and line drawing).

Table 3. Total number of microsaccades in the first second of 10 trials for each task (Task A, B and C) and subject (S–1 to S–10).

	S–1	S–2	S–3	S–4	S–5	S–6	S–7	S–8	S–9	S–10
Task A	1	0	9	3	8	1	20	0	5	1
Task B	5	0	5	2	7	1	22	8	13	0
Task C	3	3	7	0	18	1	20	24	19	4

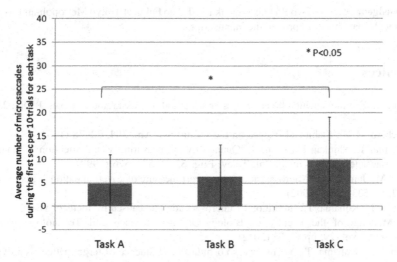

Fig. 9. Average number of microsaccades in the first second in 10 trials for each task

Table 4. Results of one-way analysis of variance using data from Table 3

Source	SS	d.f.	MS	F
Subject	1168.6667	9	129.8519	–
Task	137.4000	2	68.7000	3.32+
Subject × Task	371.9333	18	20.6630	–
Total	1678.0000	29	–	–

+p < 0.10. SS: sum of squares; d.f.: degrees of freedom; MS: mean square; F: F-ratio.

5 Conclusion and Future Work

The purpose of this study was to investigate a method for classifying the visual attention level using microsaccades measured during tasks in which a subject gazes at a visual target. In the experiment, 10 subjects performed three tasks requiring different degrees of attention. One-way analysis of variance of microsaccades performed in the first second after changing the visual target showed that the relationship approached significance ($p < 0.10$), suggesting a trend towards a positive correlation between the level of visual attention and the number of microsaccades, and multiple comparison showed that there was a significant difference between Task A and Task C ($p < 0.05$, LSD). These findings revealed that the number of microsaccades differed among the tasks. Therefore, our results suggest that visual attention level can be classified based on the number of microsaccades.

In this paper, we used visual targets requiring three different levels of attention. Future experiments should consider the use of visual targets that increase the level of attention in a linear manner. Additionally, the relationship between microsaccades and visual attention level should be confirmed using more complicated visual targets.

Acknowledgements.   We would like to thank Dr. Takao Fukui at Tokyo Metropolitan University for comments that greatly improved the manuscript.

References

1. Ukai, K.: Eye movement: characteristics and method of measurement. Japan. J. Opt. **23**, 2–8 (1994)
2. Pritchard, R.M.: Stabilized images on the retina. Sci. Am. **204**, 72–78 (1961)
3. Kohama, T., Shinkai, K., Usui, S.: Quantitatively measuring visual attention by analyzing microsaccades. J. Inst. Image Inf. Telev. Eng. **52**(4), 571–576 (1998)
4. Zaid, M., James, J.C.: Microsaccades as an overt measure of covert attention shifts. Vis. Res. **42**(22), 2533–2545 (2002)
5. Handa, T., Kohama, T.: Inhibition of microsaccade rate caused by focus of visual attention. In: Memoirs of the Faculty of Biology-Oriented Science and Technology of Kinki University, vol. 29, pp. 7–15 (2012)
6. Ohtani, S., Kohama, T., Yoshikawa., Yoshida., H.: Influence of target patterns on dynamic characteristics of fixation eye movements, IEICE Technical report, Institute of Electronics, Information and Communication Engineers, HIP2016-42-60, vol. 113, no. 229, pp. 61–66 (2016)
7. Kanoh, Y., Kohama, T., Kikkawa, S., Yoshida, H.: The relationship between spatial extent of focal attention and latencies of microsaccades. Trans. Japan. Soc. Med. Biol. Eng. **53**(O1-05-5) (2014)
8. Engbert, R., Kliegl, R.: Microsaccades uncover the orientation of covert attention. Vision. Res. **43**(9), 1035–1045 (2003)
9. Laubrock, J., Engbert, R., Kliegl, R.: Microsaccade dynamics during covert attention. Vis. Res. **45**(6), 721–730 (2005)
10. Takahashi, M.: A study on covert attention using microsaccadeas an index: introducing anticipatory response tasks toward tennis serve, Kyushu Institute of Technology University, 2018, Ph.D. thesis (2018)
11. Kashiwara, K., Okanoya, K., Kawai, N.: Effects of emotional pictures on eye movements: IEICE technical report, Human communication science, vol. 110, no. 34, pp. 41–46, HIP2010-8 (2015)

12. Sakaino, H., Tetutani, N., Kishino, F.: Spatio-temporal analysis of eye movement patterns based on the maximum entropy method and the higher moment features. IEICE Trans. Fundam. Electron. Commun. Comput. Sci. (Jpn Ed.) **76**(8), 1027–1041 (1993)
13. Carpenter, R.H.S.: Movements of the Eyes (2nd Edn. Revised and Enlarged), pp. 124–138. Pion London (1988)
14. Hoffman, J.E., Subramaniam, B.: The role of visual attention in saccadic eye movements. Percept. Psychophys. **57**(6), 787–795 (1995)
15. Deubel, H., Schneider, W.X.: Saccade target selection and object recognition: evidence for a common attentional mechanism. Vis. Res. **36**(12), 1827–1837 (1996)
16. Yoshimatsu, H.: Smoothing of miniature eye movement using wavelet analysis when gazing a fixed target. J. Inst. Telev. Eng. Jp. **50**(12), 1903–1912 (1996)
17. Noguchi, D., Kohama, T., Yoshikawa, S., Yoshida, H.: Microsaccade detection with a order-statistic low-pass differentiation filter. In: ITE Winter Annual Convention 2011, 12-2-1 (2011)
18. Emoto, J., Hirata, Y.: Realtime microsaccade detection with convolutional neural network. IEICE Trans. Fundam. Electron. Commun. Comput. Sci. **5**(2), 456–467 (2018)
19. Miyatake, K., Kohama, T., Yoshida, H.: A microsaccade detection method using a recursive neural network. In: Abstracts and Proceedings of the Annual Conference of Japanese Society for Medical and Biological Engineering, vol. 57, no. O2-7-3-2 (2018)
20. Suzuki, K., Toyoda, H., Hanayama, R., Ishii, K.: Development of binocular microsaccade measurement system using intelligent vision sensor and evaluation of synchronization between left and right microsaccades. Trans. Japan. Soc. Med. Biol. Eng. **53**(5), 99.247–99.254 (2015)
21. Fukuda, T.: The functional difference between central vision and peripheral vision in pattern perception. J. Inst. Telev. Eng. Jpn. **32**(9), 492–498 (1978)
22. Martinez-Conde, S., Macknik, S.L., Hubel, D.H.: The role of fixational eye movements in visual perception. Nat. Rev. Neurosci. **5**, 229–240 (2004)
23. Zuber, B.L., Stark, L., Cook, G.: Microsaccades and the velocity-amplitude relationship for saccadic eye movements. Science **50**, 1459–1460 (1965)
24. Ueda, T.: Soukan ga aruka wo mitsukeru kanbenhou, Operations research as a management science Communications of the Operations Research Society of Japan, vol. 42, no. 7, pp. 99.493–496 (1997)

Multiscale Analysis of Textual Content Using Eyegaze

Aniruddha Sinha[1]([✉]), Rikayan Chaki[2], Bikram Kumar De[2],
Rajlakshmi Guha[3], Sanjoy Kumar Saha[2], and Anupam Basu[4,5]

[1] TCS Research and Innovation, Tata Consultancy Services, Kolkata, India
aniruddha.s@tcs.com
[2] Department of Computer Science and Engineering, Jadavpur University,
Kolkata, India
rikayan@acm.org, bikramkumarde@yahoo.in, sks_ju@yahoo.co.in
[3] Center for Educational Technology, Indian Institute of Technology, Kharagpur,
Kharagpur, India
rajg@cet.iitkgp.ac.in
[4] Department of Computer Science and Engineering,
Indian Institute of Technology, Kharagpur, Kharagpur, India
anupambas@gmail.com
[5] National Institute of Technology, Durgapur, Durgapur, India

Abstract. Reading of a textual content involves a complex coordination between various parts of brain responsible for visual inputs, language processing, cognitive functions and motor response. In addition, psychological factors like attention and perception play a major role in understanding of the content. Many of these factors get reflected in the behaviour of eye movement, as the content is read. In this paper, we present an approach for analysing a textual content in various scales using eyegaze. The scales include (i) individual fixation characteristics, (ii) saccades and fixations within a line (iii) overall difficulty score of the content. An affordable infrared eye tracking device is used to capture the gaze characteristics in an unobtrusive manner. Two types (easy and difficult) of textual contents are designed for the experiment which are benchmarked using standard readability indices. The fixation characteristics include fixation duration, change in drift direction within a fixation and spatial area of a fixation. Using Analysis of Variance (ANOVA), the former two are found to be statistically significant in distinguishing the two types of contents. Within a line, the spatial distance between fixations and the number of switching between saccades and fixations characterize the flow during a reading where the later is found to be statistically significant. A mixture of two partial sigmoid is used as a mapping function to compute the difficulty score of a content from the significant features. For a given content, the variation of these scores among individual readers, enables us to get deeper insights into their cognitive and psychological aspects.

Keywords: Fixation · Eyegaze · ANOVA · Multiscale · Textual content

© Springer-Verlag GmbH Germany, part of Springer Nature 2020
M. L. Gavrilova et al. (Eds.): Trans. on Comput. Sci. XXXV, LNCS 11960, pp. 12–35, 2020.
https://doi.org/10.1007/978-3-662-61092-3_2

1   Introduction

Understanding of individual components in a learning material, followed by their integration in the brain, leads to the derivation of the holistic knowledge of the content [1]. Examples of these components are diagrams, graphs, tables, texts, animations etc. Various cognitive and phychological factors, play a major role in understanding of the learning material [2]. In general, the skill of an individual is defined as the ability to perform a given task. On the other hand, the difficulty level of a task is considered as the challenge experienced by an individual. Whenever there is a balance between the challenge and the skill, then an optimum performance is achieved, enabling the flow state [3]. The individual becomes anxious, if the challenge is much higher than the skill. On the contrary, when the skill level is higher than the challenge, then one experiences a state of boredom.

In the case of textual reading, a perception of the content is built through a cyclic process where the visual inputs are processed to extract the semantic information from the words and phrases [4]. Depending on the difficulty level of the text, the eye movements are effected which are controlled by cranial nerves. It reflects the demand in the cognitive load and speed in which the content is read [5]. During a task, amount of working memory used by an individual, provides a measure of the mental workload and is termed as cognitive load [2]. During a textual reading, it is believed that there are two types of control in eye movements namely, *global* and *direct* control. In case of *global* control, the overall difficulty of the text plays a dominant role, rather than individual words [6]. On the other hand, the *direct* control is more governed by individual words or phrases, which are presently processed [7].

Distance learning is gaining popularity, in order to augment the knowledge gained in the traditional learning systems. Through advancements in technology, there are continuous attempts to improve the student teacher interaction in the distance learning scenario [8]. Recently, the amount of on-line tutorials have also increased to support such learning methodologies [9]. However, the success of the overall system lies in synergy between individual capabilities and the challenge to consume these tutorials. Thus there is a need to categorize the contents in advance, in terms of their difficulty levels. This would enable to find the most suitable content, based on the capability of a reader.

Traditionally the evaluation of a content is done by subject matter experts. However, such gross evaluations, using feedback based methods, are prone to individual biases [10]. In case of textual contents, the grading is done using natural language processing based standard readability indices. These include SMOG Index, Coleman Liau Index, Flesch Kincaid Grade Level, Flesch Kincaid Reading Ease [11] etc. Additionally, statistical methods [12] measuring the density of complex words, average syllables per word are used to quantify the difficulty level of a content. Mayer [13] has studied the principles behind understanding of a multimedia content, where cognitive load [2] or mental workload is found to play a major role. Hence, there is a need to evaluate a content, by measuring the cognitive behaviours of individuals selected from the targetted audience. Additionally, personalization of contents are possible by quantifying the reading characteristics of an individual. One such means of quantification is to measure the physiological signals, as an effect of the experienced cognitive load.

During processing of a task, the collective neural responses from the brain, is a direct measure of the effect in physiology. One such means to capture the brain signal is using Electroencephalogram (EEG) [14] sensors placed on the scalp. The next direct measure is the eye movement [15], which reflects the motor neural responses from the cranial nerves. There are various indirect means of measuring the physiological responses arising due to the control of Autonomous Nervous System [16]. The indirect means include skin temperature [17], Galvanic Skin Response (GSR) [18], Electromyogram (EMG) [19], Electrocardiogram (ECG) [20], Photoplethysmogram (PPG) [21] etc. However, there is a need to wear some sensor for all the indirect means. In case of direct sensing, one also needs to wear an EEG [22] device for capturing the brain signals. Whereas, the eyegaze and pupillary responses [15] can be captured using an eye tracker, placed placed at a distance from the individual, hence avoiding any wearables.

Apart from the unobtrusive nature, eye tracking devices can capture fixations and saccades during reading of a textual content. Fixation is characterized by a relatively small amount (approximately $<2°$ in visual angle) of gaze movement, for a duration of approximately 250–300 ms [23]. In between fixations, fast movements having high rotation velocity (500–900°/s) with relatively larger amount (approximately 2–10° in visual angle) and lasting for 2–40 ms, are termed as saccades [24]. These information reflects the attention, engagement and cognitive load during learning [25] and quantifies the reading characteristics in a granular manner [15]. Finally, we emphasise the fact that the eye tracking during a silent reading, can be used to evaluate a content for a targetted group of readers, as well as characterize the reader for a content.

In this paper, we present an approach to evaluate a content in multiple scales, based on individuals reading characteristics. At the finest scale the morphology of a fixation is analysed. In the intermediate scale the relationship between fixations are evaluated for a particular textual line. Finally, at the global scale a difficulty score is generated for the overall content. Two types of textual contents [26] are chosen for the experiment, one being easy and the other difficult as benchmarked using standard readability indices. Moreover, in order to suite the mass deployment of such solutions, we have used an affordable eye tracker device , named EyeTribe [27]. To best of our knowledge, non of the previous works attempted for this type of multiscale analysis of textual content, using gaze behaviour obtained from a low cost eye tracking device. A summary on the contributions of this paper are given below:

- Approach for multiscale statistical analysis of the characteristics of fixations, between two types of contents (easy and difficult). The features used are fixation duration, change in drift direction within a fixation and spatial spread of a fixation. These features reflect the perception and attention of an individual during reading. In our previous works [26,28], we used these features in isolation and mainly for studying the formation of possible groups among individuals. In the present work, we have consolidated them as the finest level information in the multiscale analysis. Additionally, we have found the statistically significant features, responsible for discriminating between the easy and difficult contents.

- Spatio-temporal relationship between adjacent fixations within a textual line namely, spatial distance between fixations and number of switching between saccades and fixations. The switching feature used in our previous work [26] is combined with the distance feature, to do the line level analysis.
- Deriving a difficulty score for the contents by mapping the statistically significant features using a function based on mixture of partial sigmoid. Earlier, in our previous work [29], such a function was proposed and analyzed, on a different set of features. There certain tuning parameters were fixed. In the present work, we have generalized the parameters and also provided the philosophy behind selecting the tuning parameters.

The present work extends the conference paper [26]. Here a detailed analysis is presented based on the findings of the difficulty scores from all the three scales namely, fixation, line and overall content. We also demonstrate the strength of fusion of score, derived from multiple features. It helps in improving the discriminative power of the textual contents with varying difficulty. Finally, the individual scores provide insight into their language related background, reading behavior and concentration during the reading.

The organization of this paper is as follows. The review of previous works, on analysing textual contents using eyegaze, is presented in Sect. 2. The methodology of processing the eyegaze signals, in various scales namely, fixation, line and content level is given in Sect. 3. The details on the experimental setup, description on the tasks and participants, data capture protocol are presented in Sect. 4. Results are given in Sect. 5 and finally the conclusion in Sect. 6.

2   Review of Prior Works

Infrared (IR) eye trackers are quite promising in analysing the gaze behaviour, during a given task. The application areas include psychology, content analysis, usability testing, medicine, human computer interactions etc. Most of these applications use high end costly eye tracking devices. During text comprehension, *EyeLink 1000* eye tracker is used to evaluate the cognitive load by analysing the fixation characteristics [30]. Additionally, a chin rest is used for the experiment. The design principles to improve in the usability of an application, are analysed [31] by considering the effect of tone in various colors and their contrasts. Motivated by this work, Navarro et al. [32] used a costly eye tracker *Tobii X60* to analyse the effect on the eyegaze due to the highlighting of texts with various colors. During newspaper reading, gaze paths followed during the reading along with the entry and exit points for each paths, are analyzed [33] using a costly head mounted eye tracker device. *Tobii-X2-60* is used to analyse the fixation and saccades during reading of a content in second language [34]. The evaluation of reading performance [35] is presented using eye tracking devices with 500 to 1000 Hz of sampling rate.

Various psychological factors play a major role during reading of a textual content. Attention effects the understanding of the content and is also responsible for the saccadic behavior [36]. There exist several models to characterize the eye movement behavior. The saccadic behaviors, in a reading task, are predicted using E-Z Reader model [37]. Mindless reading [38] is another major factor, effecting the reading pattern of an individual. This is characterized by more fixation duration, reduced fixation rate and lesser saccade duration [39]. Due to higher cognitive load, increased fixation duration and spatial length of saccades are observed, while reading of relatively difficult texts [24]. In the area of medical applications, Forssman et al. [40] used *Tobii-X2-60*, a costly eye tracker device, to analyse the gaze behaviour on malnourished children. Burton et al. [41] and Maruta et al. [42] used a silent reading of textual content to evaluate Glaucoma patients.

It can be seen that all the studies use high-end eye tracking devices and none of them focus on the evaluation of textual content in an e-learning scenario. However, such devices are infeasible to be used for large scale deployment. Recently, affordable eye trackers are used to analyse the gaze and pupil dilation using some standard psychological test [43] for rehabilitation applications. A digitized version of a neuro-motor coordination task [44] is used to get insights into the eyegaze using a low cost eye tracker. Hence there is a need to characterize the gaze behaviour, specially during textual reading, using low cost and affordable eye trackers. Moreover, there is a need to go beyond just categorizing a content as easy or difficult. In order to get the relative ranking among the contents, it is important to derive a normalized difficulty score. Additionally, to get deeper insights into the contents, an approach for multiscale analysis is also required.

3   Methodology

Segments of texts are read in incremental manner to understand the meaning of a textual content. A set of words/phrase in a textual content is termed as a segment of text. During this processes the eyegaze is periodically localized in a segment of text for short duration and then moves to the next segment. This is done through momentary fixations, followed by saccadic movements and vice-versa. The characteristics of these fixations and saccades could give insights into the nature of the content as experienced by the reader and is a reflection of the processing done by the brain. Usually textual contents are arranged in multiple lines. In the current research work, we aim at analysing the content at multiple scales namely, fixation as the most granular level, line as an intermediate level and finally the content as a whole. Various steps of processing is shown in the block diagram in Fig. 1.

Initially, the raw eyegaze data captured from the EyeTribe is pre-processed to get rid of sensor and measurement noise. Next the spatial and temporal segments

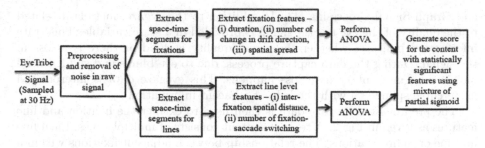

Fig. 1. Flow chart for generating a score during textual reading using multiscale analysis

of the gaze data corresponding to the fixations and line segments are extracted for the whole content. Features related to the characteristics of fixations in terms of time duration, spatial spread and change in gaze drift within a fixation are extracted. Next, the line level features are derived from the relationship between the fixations and saccades within a line. These features include the spatial distance between successive fixations and number of switching between fixations and saccades within a line. The fixation and line level features contain information related to perceptual span and attention of an individual which helps in measuring the mental workload of the reader. ANOVA analysis [45] is performed on the fixation and line level features derived from two types (easy and difficult) of textual contents, to find the statistically significant ones. Finally, the significant features are mapped to a score, using a non-linear function. The combined score from all the individuals are used to generate the overall difficulty score for the content. Moreover, the scores derived for each individual using the fixation and line levels, give information on the textual segments where one finds it difficult or easy to read.

3.1 Pre-processing of the Eye Tracker Data

An IR based eye tracker device named EyeTribe, placed in front of the subject, logs the eyegaze at a sampling rate of 30 Hz. It captures the X-Y co-ordinates of the display screen wherever the subject looks at. The sensor signal is corrupted with various internal and external noise [46]. The internal noise is mainly due to the inherent quality of the sensor which detects the eyegaze. The external noise sources are many namely, interferences due to external IR sources, eye blinks and head movements of the subject. In order to remove the effect of these noise, the preprocessing is performed on the gaze data. Initially, the inherent noise and the ambient interference on the IR are calibrated using an application provided in the Software Development Kit (SDK) by EyeTribe. A mapping is generated between the location where the patch was displayed and the eyegaze obtained by the IR sensor. After the calibration is successfully done, the actual experiment for data capture starts where the textual stimulus is display on the screen for reading. The systematic and variable noises [47] are filtered using Kalman filter and Graph Signal Processing [43]. In addition to the eygaze, metadata related to the quality of the eye tracker signal is provided in a state variable. This state information indicates whether the signal quality is good or there is a loss in the signal. During the data capture process, due to eye blinks and sudden head movements, momentary data loss happens. This eyegaze data is interpolated using cubic spline [48] with the help of adjacent information.

The preprocessed eyegaze data is then used to extract the fixation and line features as given in Fig. 2. A textual content consists of multiple lines. Each line has one or many fixations. The relationship between adjacent fixations within a line is considered as the line features. The spatio-temporal characteristics of a feature is used to derive the fixation features. These line and fixation features provide information about the difficulty in reading of the content along with reading characteristics of an individual. The pictorial representation of the lines in a content, fixations, saccades, gaze locations and scan paths are shown in Fig. 2.

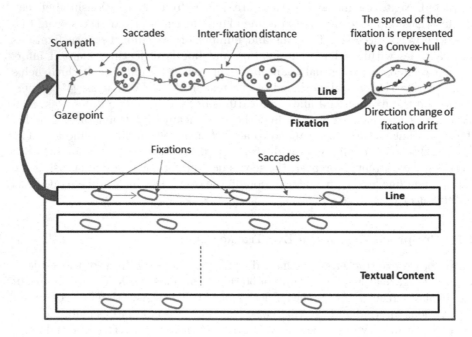

Fig. 2. Approach for the multiscale content analysis - characterisation of fixations and relationship between fixation and saccades within a line of a textual content

3.2   Extraction of Fixations Features

During the reading of textual contents, there are successive fixations and saccades. A time segment of approximately 250–300 ms [23], where the eyegaze remains relatively static, with a visual angle less than 2 degree,s is termed as a fixation. The change in the mean values of the eyegaze data, in both X and Y co-ordinates, within a window is used to detect the fixation [49]. The spatio-temporal characteristics of the fixation is used to derive the fixation features. These include (i) fixation duration, (ii) change in drift direction within a fixation and (iii) spatial area of the fixation, which are explained below.

Fixation Duration: The fixation duration feature was used in one of our previous work [26] and was normalized with line duration to compute the entropy of the content. To study the characteristics of the fixations, in the present work, we use the fixation duration feature independent of the duration of line. The start and end time stamps of a fixation are obtained using a velocity based approach [50] in the shift of mean gaze coordinates. The mean in X and Y co-ordinates are computed by considering successive gaze points. The time window in which the velocity of both the means in X and Y direction remain within a threshold, is taken as the fixation duration feature f_{dur}^i of the ith fixation. This threshold is empirically found to be 20 pixels/sample [51] using a 21 in. display with resolution 1600×1200 pixels.

Change in Drift Direction Within a Fixation: The direction change within a fixation was used as a feature for finding similarity in reading traits for individuals [28]. A brief description of the feature is given here for the sake of completeness. Within the time window of a fixation, the eyegaze drifts spatially with very small amount and the direction of the drift also changes. This is due to the neural control done by the brain to stabilise the eye muscles. However, due to attention [36] there is a change in drift direction during a fixation. In the right top corner of Fig. 2, a sample fixation depicts how the gaze points changes with time. The red arrow connects the gaze points from left to right and then the blue arrow indicates the movement of gaze points from right to left, indicating a change in drift direction. Finally, there is another change in direction indicated by the red line. Thus in this example the change in direction happens twice. For the ith fixation, this number of change in drift direction is taken as a feature f_{ncd}^i.

Spatial Area of Fixation: The spatial spread of fixation was used as a feature in our previous work for grouping individuals based on their reading characteristics [28]. The feature is described briefly for the ready reference for the readers. Within a fixation the gaze points slowly drifts below a threshold velocity and covers a spatial region in the X and Y coordinates. The convex hull thus formed by the gaze points indicates the spatial span of the fixation. In the right top corner of the Fig. 2, the spatial boundary shown for the fixation denotes the convex hull. For the ith fixation, the spatial area of the convex hull f_{sac}^i is taken as the feature. This reflects the perceptual span [52] during a fixation.

3.3   Extraction of Line Feature

While reading a textual content in English, as a line is read, the eyegaze moves gradually from left to right of the screen. Thus the X coordinate of the gaze point increases while the line is read and the Y coordinate remains relatively constant. At the end of the line, the gaze suddenly moves to the beginning of the next line leading to a sudden decrease in X coordinate and small amount of increase in Y coordinate. This leads to a oscillatory nature of the gaze data in X coordinate where a cycle corresponds to a line read in the text [26]. The troughs [53] and peaks correspond to the start and end of the lines respectively. Once the time segments corresponding to the lines are detected, the fixations within a line are considered to derive features which corresponds to the relationship between fixations within a line. These are the spatial distance between adjacent fixations and number of alternate fixations and saccades within a line.

Inter-Fixation Spatial Distance: Within a line there could be multiple fixations as shown in Fig. 2. The distance between the successive fixations are considered as the features. Initially the centroids of the fixations are computed using the gaze points within respective fixations. The pixel distance between the centroids of the adjacent fixations are taken as the inter-fixation spatial distance. The mean, a primitive statistical parameter, of these distances are taken to derive the feature f_{fsd}^l corresponding to l^{th} line.

Number of Switching Between Fixation and Saccades: While reading the fixations and saccades occur alternately. The number of times this transition happens between the two is used to derive the switching feature. The same feature was used in one of our previous work [26] to derive the entropy of the content. As the time taken by different readers vary, the number switching is normalized with the duration taken to read the line. This normalized value is taken as the features f_{nns}^l for the lth line.

Each of these fixation and line features are derived from the gaze data captured from all the subjects or individuals to create the feature sets. The fixation feature set is $f_f = \{f_{dur}^{i,s}, f_{ncd}^{i,s}, f_{sac}^{i,s} : 1 \leq i \leq N_S, 1 \leq s \leq S\}$ and the line feature set is $f_l = \{f_{fsd}^{l,s}, f_{nns}^{l,s} : 1 \leq l \leq L_S, 1 \leq s \leq S\}$. Here the N_S and L_S corresponds to the number of fixations and number of lines for the s_{th} individual respectively and S is the total number of individuals. Next for the fixation and line features, two subsets of features are created, one corresponding to while reading the easy content $(f_{f,easy}, f_{l,easy})$ and the other for the difficult content $(f_{f,diff}, f_{l,diff})$. We consider the NULL hypothesis as the fixation features $f_{f,easy}$ and $f_{f,diff}$ are similar. ANOVA analysis is performed on individual features to find the statistical difference between the two set of features. The NULL hypothesis is rejected if the p value of the ANOVA is less than 0.05. Thus the statistically significant fixation features are derived for which the NULL hypothesis is rejected. Similar analysis is done for the line features. The significant fixation and line features are used to compute the difficulty score of each features. The scores derived from the individual features provide two types of information namely, aggregate information of the overall content and the insights into the segments of texts where individuals find it relatively difficult. The aggregate information of a contents (easy or difficult) is derived from the average values of the scores obtained from all the individuals. More importantly, the fixation locations and the lines having higher scores provide insights into the texts corresponding to them.

3.4   Generation of Difficulty Score

In order to obtain a difficulty score from the significant features, a normalized mapping function is needed. Such a function ($S\colon \mathbb{R}^+ \to [0,1]$) was proposed in our previous work [29] where certain tuning parameters were fixed. In the philosophycurrent work we have generalized the parameters. We briefly describe the mapping function which is based on mixture of partial sigmoid and the fusion of such mapping functions derived from multiple features. The trade-off between a good dynamic range and discriminative capability is discussed in the design of the parameter section.

The continuous function maps a feature (f) from a real number $\mathbb{R}$ to the range [0 1] in order to generate the normalized difficulty score. Here the 0 indicates least difficult and 1 indicates most difficult. The design of the mapping function needs to be such that there is a good trade off between the separability of the low and high feature values while maintaining a good dynamic range. This indicates that the function needs to be non-linear as well as should be configurable based on the distribution of the features. A sigmoid function S as given in (1) meets such criterion. Moreover, it is of special interest as the slope of these functions increase monotonically till a point f_0 beyond which it starts deceasing. The gain parameter g of the function can be configured to adjust the dynamic range.

$$S(f) = \frac{1}{1 + e^{-g(f-f_0)}} \tag{1}$$

In the present work we have two types of contents namely, easy and difficult. Hence it is expected that the features would be distributed on either side of the center value. However, the distribution of the features corresponding to the easy and difficult texts might be different. Hence the mapping function should have atleast two configurable parameter corresponding to the easy and difficult levels of the content. This is achieved by constructing a mixture of partial sigmoid.

Mapping Function - Mixture of Partial Sigmoid: A standard sigmoid function is symmetric around the center point (0.5). We use two partial sigmoids (one from each half of the symmetry point) with different g values to construct the mapping function. This approach enables us with the freedom to tune the function based on the feature distributions obtained from the easy and difficult texts. A sample illustration of the mapping from the feature distribution to the mapping function is given in Fig. 3.

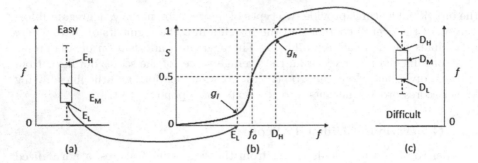

Fig. 3. (a) Sample feature distribution (f) for easy content represented in a Box-plot, (b) Example of the function (S) with mixture of partial sigmoid, (c) Sample feature distribution (f) for difficult content represented in a Box-plot

A box plots of the features are shown in Fig. 3(a) and (c) where the E_M and D_M are the median values of the two set of features respectively. Next a quantile analysis is done on each of the feature set. The E_L, D_L are the quantile values for the low percentile of the easy and difficult features respectively. Similarly, E_H, D_H correspond for high quantiles. These parameters are used to generate the mapping function S as shown in Fig. 3(b). The center point f_0 is taken as the average of two median values. The gain parameters are g_l and g_h for the two partial sigmoids $S_l(f)$ and $S_h(f)$ corresponding to the lower and higher values of the f_0 respectively as given in (2). The final continuous mapping function $S(f)$ the sum of the individual ones with the value of 0.5 at $f = f_0$.

$$S_l(f) = \frac{1}{1 + e^{-g_l(f-f_0)}} \quad \forall f < f_0$$
$$= 0 \quad otherwise$$
$$S_h(f) = \frac{1}{1 + e^{-g_h(f-f_0)}} \quad \forall f > f_0$$
$$= 0 \quad otherwise$$
$$S(f) = S_l(f) + S_h(f)$$
$$= 0.5 \qquad at \ f = f_0 \tag{2}$$

Next we present the methodology to find the gain parameters g_l and g_h by selecting the quantiles E_L and D_H. It is to be noted that for further analysis using $S(f)$ we would be performing differentiation on individual functions namely, $S_l(f)$ and $S_h(f)$ and hence $S(f)$ being non-differentiable at $f = f_0$ is not a hindrance in computation.

Design of Parameters: In order to maintain a balance between the separation of the features between two types of content as well as good dynamic range, we need to tune the gain parameters for the mapping function S. The feature distributions for two types of content are non-uniform and could be quite different. As mentioned earlier, the center point (f_0) is chosen as the average of the medians $(E_M$ and $D_M)$. The gain parameters g_l and g_h are tuned in such a manner that the inflexion points of the first derivative of the partial sigmoids match the low (E_L) and high (D_H) quantile values. In the Fig. 4 we depict sample sigmoid functions and their first and second derivatives to illustrate the design of the gain function. As the gain parameter of the sigmoid function (1st row in Fig. 4) is changed, the inflexion points in the first derivative (2nd row in Fig. 4) shift which get reflected in the peaks of the second derivative (3rd row Fig. 4). These inflection points are matched with the (E_L) and high (D_H) quantile values to select the gain parameters.

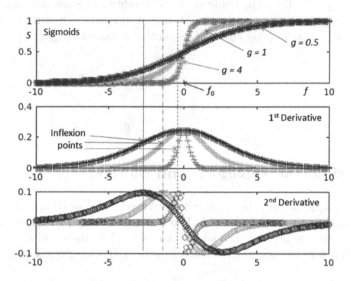

Fig. 4. Example plots for sigmoid functions are shown for different gain (g) parameter along with its 1st derivative and 2nd derivative. The change in the inflexion point for different gain parameter are shown.

From the first row of the Fig. 4, it can be seen that as the gain value increases (e.g. $g = 1$), the sigmoid becomes more steep in the proximity of the center point. The higher gain parameter helps in larger separation in the scores derived from easy and difficult features but the dynamic range is reduced. Hence the finer changes of the scores within a type of content is lost. On the other hand lower values of gain (e.g. $g = 0.1$) increases the dynamic range by the discriminative capability is reduced. For the partial sigmoid S_l, the g_l increases if E_L is increased. On contrary, for S_h, the g_h increases if D_H is reduced. Thus there is a balance required between the two. In the present work, the balance is achieved by empirically setting the values of E_L and D_H as 25 and 75 percentile respectively as quantile values.

It is important to note that the score derived by this process is named a difficulty score. This means that for difficult texts this score is expected to be higher. In other words, if E_M is lower than D_M then the score derived using the mapping function truely gives the difficulty score. However, due to the nature of the features it may so happen that the E_M is less than D_M. In such a scenario, the output of the mapping function is subtracted from the maximum value (which is 1) and thus the difficulty score is obtained.

Fusion of Score: The fusion of score can be done for the scores obtained from the statistically significant (p value < 0.05 in ANOVA) features within the same scale (namely, fixation or line level). Each features would have their own discriminative power which gets reflected in the F-value of ANOVA. In order to fuse multiple scores from the same scale, a weighted average is taken between them. The weight for the score (S^i) obtained from the ith feature is w_i. These weights are set in proportion to their respective F-values ($F_{value}(i)$).

$$FusedScore = \frac{\sum_{i=1}^{N} w_i \times S^i}{\sum_{i=1}^{N} w_i} \tag{3}$$

4   Experimental Apparatus and Method

In the present experiment, the eyegaze is captured using an IR based eyetracker, while a textual content is being read. The textual contents, setup for the experiment, data capture protocol are similar to as used in the previous works [26], [29]. However, for the continuity, we are briefly elaborating the same.

4.1   Textual Content as Stimulus

The stimuli used for the experiment are multiple paragraphs of textual contents. The task is to silently read those contents. In that process the eye movement behaviour is captured using the IR based eye tracker. The objective is to derive features of eyegaze using fixations and saccades and quantify the effect based on varying difficulty levels. Thus two types of textual contents are used, such that one is easy to read and comprehend whereas the other is difficult to read. For each of the easy and difficult types, three paragraphs are created. The paragraphs for the easy contents are designed using selected sample texts[1] used for learning English by kids. One paragraph for the difficult content is created from snippets of the novel Great Expectation[2]. The other two difficult paragraphs are created

---

[1] http://www.preservearticles.com/2011080510137/12-short-paragraphs-in-english-language-for-school-kids-free-to-read.html.

[2] https://www.enotes.com/topics/great-expectations.

using selected texts from internet[3]. Each of the six paragraphs contain 132–164 words and 12–14 lines.

Initially these paragraphs are benchmarked for their reading difficulty using few readability indices [11]. These indices are given below:

- Flesch Kincaid Grade Level - The values are greater than 3. Higher values indicate more difficulty level.
- Flesch Kincaid Reading Ease - The range of value is 1 to 100. Lower values indicate more difficulty level.
- Coleman Liau Index - The values are greater than 1. Higher values indicate more difficulty level.
- SMOG Index - The range of value is 4 to 18. Higher values indicate more difficulty level.

The mean values of Flesch Kincaid Reading Ease, Flesch Kincaid Grade Level, SMOG Index and Coleman Liau Index for the easy(difficult) paragraphs are 83.5(37.1), 4.7(14.6), 5.2(12.4) and 8.3(13.8) respectively [26]. Additionally, the mean values for percentage of complex words, average words in each sentence, and average syllables in each word [12] for easy(difficult) paragraphs are 6.2(18.6), 12.1(25), 1.3(1.8) respectively. It is to be noted that the average number of words and lines are similar for two types of texts, as mentioned above. Thus the separation in the difficulty levels for two types of contents are justified.

4.2   Experimental Setup

The experimental setup consists of a 21 in. display screen, IR eye tracker named EyeTribe [27] and a keyboard for user response as shown in Fig. 5. The display is placed in front of the subject at an approximate distance of 24 in.. The height of the subject is adjusted using a height adjustable chair such that the individual can comfortably read the content on the screen. is The EyeTribe is placed below the screen facing towards the subject. The placement of the eyetracker is adjusted such that both the eye are properly visible by the sensor. This is done in guidance with the SDK available with the EyeTribe. Initially, a calibration is done for the mapping between the eyegaze and the screen co-ordinates of the display. During this calibration process a circular patch moves on the display screen while momentarily being fixed in certain locations. The subject needs to follow the patch during the calibration process. There are two types of calibration namely, 9 or 16 points where the numbers indicate the number of locations where the patch gets fixed during the process. At the end of the calibration a score is generated by the SDK, between 1 to 5, indicating the quality of calibration. Higher the number, better is the quality. For our experiment we have made sure that for all the subjects, the calibration score is achieved as 5 by adjusting the setup and controlling the external IR interferences. A 9 point calibration is found to be good for us. Once the calibration is successfully done, the data collection starts following the data capture protocol.

---

[3] https://www.quora.com/Why-is-there-so-much-criticism-for-demonetization-in-India-when-it-is-undoubtedly-a-correct-step.

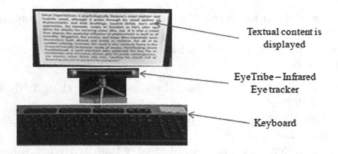

Fig. 5. Experimental setup

4.3 Data Capture Protocol

Initially every participant is given a demonstration on the method of experiment. They were asked to adjust the height the of the chair so that they are comfortable reading the texts on the screen. For this purpose they were shown different texts but of same font and layout. Once the participant is ready then the eye tracker is calibrated as mentioned in previous section. This protocol for the data capture is shown in Fig. 6. After the calibration a fixation '+' is shown on the screen and the participants are asked to look at that for 15 s. This is to create a baseline for the fixation characteristics of the subject. After that one after another text paragraphs are read for the same type (easy or difficult) of content. Once a paragraph is completely ready, the participants press the spacebar on the keyboard to move to the next paragraph. After all the three paragraphs are read for the same type of content, similar process is followed for the other type. In order to bring the participant to a baseline condition, a 3–4 minutes rest is given between the reading of two types of texts. The sequence of the two types are randomized between the participants to get rid of any bias. At the end of the experiment a questionnaire based feedback is taken on 5-point Likert scale [54] on the experienced difficulty level, their concentration level and reading habits of English novels.

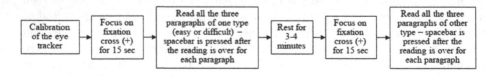

Fig. 6. The sequence of the stimulus

4.4 Participants

We have selected twenty six subjects (15 males, 11 females) having Engineering background between age range of 21–44 years. The subjects have similar cultural background and educational qualification. All the subjects have normal or corrected to normal vision. The clearance on ethical issues for handling and analysis of the data collected has been acquired from Institutional Review Board (IRB) of Tata Consultancy services ltd. We have anonymized the data for the purpose of analysis. We have followed The Helsinki Human Research guidelines (https://www.helsinki.fi/en/research/research-environment/research-ethics) for the process of data collection and handing of the same. All the participants were made aware of the experiment and an informed consent is taken from them.

5   Results and Discussion

While a textual content is read line by line, a gaze fixation enables the visual input of a segment of text which are processed by brain, after which the gaze point moves to the next segment with the help of a saccade. During this process, depending on the contents of the textual segments the fixation characteristic change. These are captured using few fixation features. Additionally, line features are extracted using the relationship between adjacent features within a line. ANOVA analysis of these features are done to find the statistically significant features. Difficulty score is derived for the significant features. The scores derived from the fixation and line features enable us to get insights at different scales of the text. It provides information about the texts corresponding to the fixations with high scores which are in agreement with most of the individuals. Similarly, the scores from the line features provide information on the difficult lines. Finally, the aggregate scores derived from all the individuals provide the difficulty level of the content. In addition to this, individual level analysis is also done to study the differences between individuals in the reading characteristics.

5.1   Scores Using Fixation Features

We have considered one temporal feature namely, fixation duration (f^i_{dur}) and two morphological features namely, number of change in drift direction (f^i_{ncd}) and spatial area of the convex hull (f^i_{sac}) formed by the gaze points within the ith fixation. Here i is the fixation index. These features are computed for all the fixations and all the individuals for both types (easy and difficult) of contents. ANOVA analysis is done on the individual fixation features extracted from these two types of content. Table 1 shows the p and F values of the ANOVA analysis, where it can be seen that f^i_{ncd} and f^i_{dur} are statistically significant in discriminating two types of contents. Hence these two are further used for generating the scores. It is to be noted that the median of both the features for the easy content is lower than the difficult content. This indicates that the fixations for the easy contents are of lower in duration compared to the difficult ones. This is in agreement with the findings made by Rayner et al. [24]. Similarly the fixations for easy contents are having less number of change in drift direction as compared to difficult contents. This indicates that the attention is more, during reading an easy content than the difficult content.

Table 1. The p and F values of the ANOVA analysis for the fixation features

Fixation features	p values	F values	Significant
f_{ncd}^i	3.26×10^{-4}	12.93	Yes
f_{dur}^i	1.49×10^{-4}	14.41	Yes
f_{sac}^i	0.89	0.02	No

In order to obtain the normalized scores for each of these features, mapping functions for f_{ncd}^i and f_{dur}^i are derived. During this the quantile values of 25th and 75th percentile for the feature distributions are used which are obtained from easy and difficult contents respectively. The f_0 is found to be 1.5 and 11.5 for f_{ncd}^i and f_{dur}^i respectively. The mapping function for the fixation duration feature (f_{dur}^i) is shown in Fig. 7 where the normalized scores are generated from the feature points. Here the blue segment corresponds to the partial sigmoid obtained from the distribution of the features taken from gaze data of easy content and the red segment for the gaze data from difficult content.

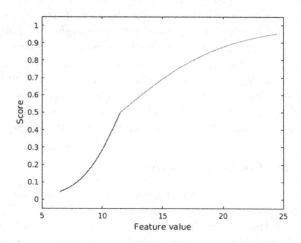

Fig. 7. Partial sigmoid based mapping function for fixation duration feature (f_{dur}^i)

A composite score is also derived using these two features where the weights for the fusion are 0.473 and 0.527 for f_{ncd}^i and f_{dur}^i respectively. These weights are obtained using their F values as described in Sect. 3.4. Next to analyse the efficacy of the score, we again perform ANOVA analysis of the scores obtained from the individual features and the composite score. The box plot of the ANOVA analysis is shown in Fig. 8.

Table 2 gives the p and F values obtained from the ANOVA analysis of individual and fused scores. It can be seen that for f_{ncd}^i, the p and F values derived from the scores are similar to that of the features. In case of the feature f_{dur}^i, the

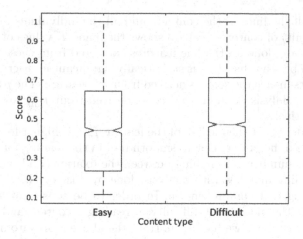

Fig. 8. The Box plot of the fused score derived from f_{ncd}^i and f_{dur}^i is shown for easy and difficult contents

scores have further improved the p and F values indicating that the mapping function is further able to improve the discrimination between easy and difficult content. In the Table 2, the improvements are indicated in bold. The fusion of these two scores shows further improvements in the p and F values, which is much better compared to the individual ones. Hence the usefulness of the fusion is demonstrated.

Table 2. The p and F values of the ANOVA analysis for the fixation scores

Fixation scores	p values	F values	Significant	Remarks
Stand-alone using f_{ncd}^i	3.94×10^{-4}	12.58	Yes	The p and F values derived from score are quite similar to the features
Stand-alone using f_{dur}^i	$\mathbf{0.26 \times 10^{-4}}$	**17.70**	Yes	Score improved both p and F values
Fusion between f_{ncd}^i and f_{dur}^i	$\mathbf{0.0195 \times 10^{-4}}$	**22.70**	Yes	The fused score further improved both p and F values

5.2   Scores Using Line Feature

The relationship between adjacent fixations are used to derive line level features. These features are inter-fixation spatial distance (f_{fsd}^l) and number of switching (f_{nns}^l) between fixations and saccades corresponding to l^{th} line. The feature f_{nns}^l is normalized by the duration which one takes to read the line. These features are computed for all the lines in the contents and all the individuals for both types (easy and difficult) of contents. Table 3 shows the p and F values of the ANOVA analysis, which are done on the line features extracted from these two types of contents. It can be seen that f^l_{nns} is statistically significant in discriminating two types of contents, hence the score is derived from this feature. The p and F values of the ANOVA analysis done on the scores derived from f^l_{nns} are 2.62×10^{-4} and 13.4 respectively.

It is to be noted that the median of the feature f^l_{nns} is higher for easy content compared to the difficult one. Due to smooth flow in the reading for easy content there is a higher number of switching between the fixations and saccades within a line. This finding matches with analysis done by Just et al. [15] on the flow of reading during text comprehension. In order to be consistent with the fact that the score is the measure of difficulty in reading a content and lower values indicate an easy content, we had to subtract the above scores from 1.

Table 3. The p and F values of the ANOVA analysis for the line features and the score

Line Features	p values	F values	Significant
f^l_{nns}	6.02×10^{-4}	11.82	Yes
f^i_{fsd}	0.37	0.80	No

5.3   Score of the Content

The scores of the contents are derived separately from the fixation and line features by taking average of all the three contents of same type (easy or difficult) for all the individuals. Table 4 shows the scores for both types of contents (easy and difficult). It can be seen that though average scores for difficult contents are higher than that of the easy content, the differences between the scores of two types of contents are apparently not much. However, there exist statistically significant differences in the two types of contents for all the three features shown in Table 4. There are many reasons behind the low difference in scores. These significant fixation and line features derived from the gaze data reflect the attention and flow of the reading. These cognitive and psychological parameters depend on individuals background and state of the mind during reading. Based on the feedback from the individuals, it is found that 60% of them were able to fully concentrate during the reading. Moreover, 14 individuals out of 26 had a background with English as first language and many of them read English novels regularly. In such a scenario the difference between the easy and difficult content is expected to be low. In order to get further insights into the individual reading traits, we analyse the individual scores.

Table 4. The scores of the contents using the significant fixation and line features

Significant feature	Easy content	Difficult content
f^i_{ncd}	0.458	0.491
f^i_{dur}	0.475	0.515
f^l_{nns}	0.466	0.527

5.4 Analysis on Individual Scores

Individual level analysis is done using the composite score derived from the significant fixation features. The similar analysis can be done using other features too. However, for demonstrating the usefulness of this analysis we have restricted ourselves only to the composite score. The range of scores for both the contents lye between 0.31 and 0.64. Out of 26 participants, the scores for 10 individuals are such that they found both the contents relatively easy. For them the range of scores are 0.35 ± 0.05 and 0.44 ± 0.04 respectively. Out of these 10 individuals, 9 of them read English novels regularly and all of them had first language as English during their school. Two individuals found both the contents relatively difficult having the range of scores between 0.59 to 0.64. Both of these individuals had English as second language and they don't read English novels. There are 8 individuals who found both the contents as average in difficulty level where the scores range between 0.47 and 0.54. Out of these 8 individuals, 6 of them mentioned that they were not able to fully concentrate during the reading. Remaining 6 individuals showed a large difference in the score between the easy and difficult contents, where the lows are in the range 0.33 ± 0.02 and the highs are in the range of 0.59 ± 0.02. All these 6 individuals has English as 2nd language in their schools. Thus it can be seen that though the scores in the content level given in Table 4 are close between the easy and difficult contents, the detailed information for the individuals give further insights into their reading characteristics.

6 Conclusion

A methodology is presented to objectively measure the difficulty level experienced during textual reading. Cognitive and psychological factors effect the reading characteristics of individuals. Eyegaze is one of the direct physiological response that can provide insights into cognitive load, attention, perceptual span and flow of reading. Multiscale approach is presented to analyse the content starting from individual fixation characteristics, spatio-temporal relationship between fixations within a line and finally scoring the overall content. Two types of contents namely, easy and difficult are used for the experiment, where three paragraphs are taken for each type. These contents are initially benchmarked for their difference in the difficulty levels using standard readability indices. ANOVA analysis of fixation and line level features are done to identify the statistically significant ones. The fixation duration and number of change in drift direction are found to be significant fixation features. The normalized number of switching between fixations and saccades within a line is also found to be significant. A mapping function based on mixture of partial sigmoid is used to derive a difficulty score of a content which helps in finer analysis of the content rather than just an easy or difficult level. Fusion of scores leads to an increase in the discrimination power between the two types of contents. Individual scores provide insight into their language related background, reading behaviour and concentration during the reading. In future, We plan to extend the present work to check the performance of the text difficulty prediction of an unseen text. Additionally, we would like to do a joint analysis of the cognitive load derived from EEG signals and its covariation with the fixation or the line level score. We also intend to perform the experiment on subject specific contents (e.g. Physics, Chemistry, Geography etc.) and use on a larger population.

Acknowledegment. Authors would like to thank the participants for their cooperation during the experiment and data collection for the same.

References

1. Demb, J.B., Boynton, G.M., Heeger, D.J.: Brain activity in visual cortex predicts individual differences in reading performance. Proc. Nat. Acad. Sci. **94**(24), 13363–13366 (1997)
2. Sweller, J.: Cognitive load during problem solving: effects on learning. Cogn. Sci. **12**(2), 257–285 (1988)
3. Czisikszentmihalyi, M.: Flow-the psychology of optimal experience (1990)
4. McConkie, G.W., Rayner, K.: The span of the effective stimulus during a fixation in reading. Percept. Psychophys. **17**(6), 578–586 (1975)
5. Carpenter, R.H.: Eye Movements. Vision and Visual Dysfunction, vol. 8. Nature Publishing Group, London (1991)
6. O'regan, J., Lévy-Schoen, A., Pynte, J., Brugaillère, B.: Convenient fixation location within isolated words of different length and structure. J. Exp. Psychol.: Hum. Percept. Perform. **10**(2), 250 (1984)
7. Just, M.A., Carpenter, P.A.: The Psychology of Reading and Language Comprehension. Allyn & Bacon, Boston (1987)
8. Mehlenbacher, B., et al.: Usable e-learning: a conceptual model for evaluation and design. In: 11th Proceedings of HCI International, vol. 2005. Citeseer (2005)
9. Raish, V., Behler, A.: Library connection: an interactive, personalized orientation for online students. J. Libr. Inform. Serv. Dist. Learn. **13**(1–2), 129–149 (2019)
10. Campbell, J.D., Tesser, A.: Motivational interpretations of hindsight bias: an individual difference analysis. J. Pers. **51**(4), 605–620 (1983)
11. Kincaid, J.P., Fishburne Jr., R.P., Rogers, R.L., Chissom, B.S.: Derivation of new readability formulas (automated readability index, fog count and flesch reading ease formula) for navy enlisted personnel. Technical report, Naval Technical Training Command Millington TN Research Branch (1975)
12. Stemler, S.: An overview of content analysis. Pract. Assess. Res. Eval. **7**(17), 137–146 (2001)

13. Mayer, R., Mayer, R.E.: The Cambridge Handbook of Multimedia Learning. Cambridge University Press, Cambridge (2005)
14. Gavas, R., Das, R., Das, P., Chatterjee, D., Sinha, A.: Inactive-state recognition from EEG signals and its application in cognitive load computation. In: 2016 IEEE International Conference on Systems, Man, and Cybernetics (SMC), pp. 003606–003611. IEEE (2016)
15. Just, M.A., Carpenter, P.A.: A theory of reading: from eye fixations to comprehension. Psychol. Rev. **87**(4), 329 (1980)
16. Loewy, A.D., Spyer, K.M.: Central Regulation of Autonomic Functions. Oxford University Press, Oxford (1990)
17. Zhai, J., Barreto, A.: Stress detection in computer users based on digital signal processing of noninvasive physiological variables. In: 28th Annual International Conference of the IEEE Engineering in Medicine and Biology Society, EMBS 2006, pp. 1355–1358. IEEE (2006)
18. Shi, Y., Ruiz, N., Taib, R., Choi, E., Chen, F.: Galvanic skin response (GSR) as an index of cognitive load. In: CHI 2007 Extended Abstracts on Human Factors in Computing Systems, pp. 2651–2656. ACM (2007)
19. McGuigan, F.J., Rodier, W.I.: Effects of auditory stimulation on covert oral behavior during silent reading. J. Exp. Psychol. **76**(4p1), 649 (1968)
20. Ryu, K., Myung, R.: Evaluation of mental workload with a combined measure based on physiological indices during a dual task of tracking and mental arithmetic. Int. J. Ind. Ergon. **35**(11), 991–1009 (2005)
21. Hjemdahl, P., Freyschuss, U., Juhlin-Dannfelt, A., Linde, B.: Differentiated sympathetic activation during mental stress evoked by the stroop test. Acta Physiol. Scand. Suppl. **527**, 25–29 (1984)
22. Khurana, V., Kumar, P., Saini, R., Roy, P.P.: Eeg based word familiarity using features and frequency bands combination. Cogn. Syst. Res. **49**, 33–48 (2018)
23. Hahn, M., Keller, F.: Modeling human reading with neural attention. arXiv preprint arXiv:1608.05604 (2016)
24. Rayner, K.: Eye movements in reading and information processing: 20 years of research. Psychol. Bull. **124**(3), 372 (1998)
25. Paas, F., Tuovinen, J.E., Tabbers, H., Van Gerven, P.W.: Cognitive load measurement as a means to advance cognitive load theory. Educ. Psychol. **38**(1), 63–71 (2003)
26. Sinha, A., Chaki, R., De Kumar, B., Saha, S.K.: Readability analysis of textual content using eye tracking. In: Chaki, R., Cortesi, A., Saeed, K., Chaki, N. (eds.) Advanced Computing and Systems for Security. AISC, vol. 897, pp. 73–88. Springer, Singapore (2019). https://doi.org/10.1007/978-981-13-3250-0_6
27. Popelka, S., Stachoň, Z., Šašinka, Č., Doležalová, J.: Eyetribe tracker data accuracy evaluation and its interconnection with hypothesis software for cartographic purposes. Comput. Intell. Neurosci. **2016**, 20 (2016)
28. Sinha, A., Kumar Saha, S., Basu, A.: Determining perceptual similarity among readers based on eyegaze dynamics. In: Chaki, R., Cortesi, A., Saeed, K., Chaki, N. (eds.) Advanced Computing and Systems for Security. AISC, vol. 996, pp. 113–124. Springer, Singapore (2020). https://doi.org/10.1007/978-981-13-8969-6_7. In Press
29. Sinha, A., Kumar Saha, S., Basu, A.: Assessment of reading material with flow of eyegaze using low-cost eye tracker. In: Das, A.K., Nayak, J., Naik, B., Pati, S.K., Pelusi, D. (eds.) Computational Intelligence in Pattern Recognition. AISC, vol. 999, pp. 497–508. Springer, Singapore (2020). https://doi.org/10.1007/978-981-13-9042-5_42. In Press

30. Raney, G.E., Campbell, S.J., Bovee, J.C.: Using eye movements to evaluate the cognitive processes involved in text comprehension. J. Vis. Exp.: JoVE **83**, e50780 (2014)
31. Johnson, J.: Designing with the Mind in Mind: Simple Guide to Understanding User Interface Design Guidelines. Elsevier, Amsterdam (2013)
32. Navarro, O., Molina, A.I., Lacruz, M., Ortega, M.: Evaluation of multimedia educational materials using eye tracking. Proc.-Soc. Behav. Sci. **197**, 2236–2243 (2015)
33. Holsanova, J., Rahm, H., Holmqvist, K.: Entry points and reading paths on newspaper spreads: comparing a semiotic analysis with eye-tracking measurements. Vis. Commun. **5**(1), 65–93 (2006)
34. Indrarathne, B., Kormos, J.: The role of working memory in processing L2 input: insights from eye-tracking. Bilingualism: Lang. Cogn. **21**(2), 355–374 (2018)
35. Kliegl, R., Laubrock, J.: Eye-movement tracking during reading. Research Methods in Psycholinguistics and the Neurobiology of Language: A Practical Guide, pp. 68–88 (2017)
36. Deubel, H., O'Regan, K., Radach, R., et al.: Attention, information processing and eye movement control. Read. Percept. Process 355–374 (2000)
37. Reichle, E.D., Pollatsek, A., Fisher, D.L., Rayner, K.: Toward a model of eye movement control in reading. Psychol. Rev. **105**(1), 125 (1998)
38. Feng, S., D'Mello, S., Graesser, A.C.: Mind wandering while reading easy and difficult texts. Psychonom. Bull. Rev. **20**(3), 586–592 (2013)
39. Reichle, E.D., Reineberg, A.E., Schooler, J.W.: Eye movements during mindless reading. Psychol. Sci. **21**(9), 1300–1310 (2010)
40. Forssman, L., et al.: Eye-tracking-based assessment of cognitive function in low-resource settings. Archives of Disease in Childhood, archdischild-2016 (2016)
41. Burton, R., Saunders, L.J., Crabb, D.P.: Areas of the visual field important during reading in patients with glaucoma. Jpn. J. Ophthalmol. **59**(2), 94–102 (2015)
42. Murata, N., Miyamoto, D., Togano, T., Fukuchi, T.: Evaluating silent reading performance with an eye tracking system in patients with glaucoma. PLoS One **12**(1), e0170230 (2017)
43. Gavas, R.D., Roy, S., Chatterjee, D., Tripathy, S.R., Chakravarty, K., Sinha, A.: Enhancing the usability of low-cost eye trackers for rehabilitation applications. PLoS One **13**(6), e0196348 (2018)
44. Chatterjee, D., Gavas, R.D., Chakravarty, K., Sinha, A., Lahiri, U.: Eye movements-an early marker of cognitive dysfunctions. In: 2018 40th Annual International Conference of the IEEE Engineering in Medicine and Biology Society (EMBC), pp. 4012–4016. IEEE (2018)
45. Keselman, H., et al.: Statistical practices of educational researchers: an analysis of their ANOVA, MANOVA, and ANCOVA analyses. Rev. Educ. Res. **68**(3), 350–386 (1998)
46. Johansen, S.A., San Agustin, J., Skovsgaard, H., Hansen, J.P., Tall, M.: Low cost vs. high-end eye tracking for usability testing. In: CHI 2011 Extended Abstracts on Human Factors in Computing Systems, pp. 1177–1182. ACM (2011)
47. Hornof, A.J., Halverson, T.: Cleaning up systematic error in eye-tracking data by using required fixation locations. Behav. Res. Methods Instr. Comput. **34**(4), 592–604 (2002)
48. De Boor, C., De Boor, C., Mathématicien, E.U., De Boor, C., De Boor, C.: A Practical guide to Splines, vol. 27. Springer, New York (1978)
49. Veneri, G., Federighi, P., Rosini, F., Federico, A., Rufa, A.: Influences of data filtering on human-computer interaction by gaze-contingent display and eye-tracking applications. Comput. Hum. Behav. **26**(6), 1555–1563 (2010)

50. Salvucci, D.D., Goldberg, J.H.: Identifying fixations and saccades in eye-tracking protocols. In: Proceedings of the 2000 Symposium on Eye Tracking Research and Applications, pp. 71–78. ACM (2000)
51. Sen, T., Megaw, T.: The effects of task variables and prolonged performance on saccadic eye movement parameters. In: Advances in Psychology. vol. 22, pp. 103–111. Elsevier (1984)
52. Rayner, K., Well, A.D., Pollatsek, A.: Asymmetry of the effective visual field in reading. Percept. Psychophys. **27**(6), 537–544 (1980)
53. Jacobson, A.: Auto-threshold peak detection in physiological signals. In: Engineering in Medicine and Biology Society, Proceedings of the 23rd Annual International Conference of the IEEE. vol. 3, pp. 2194–2195. IEEE (2001)
54. Hinkin, T.R.: A brief tutorial on the development of measures for use in survey questionnaires. Organ. Res. Methods **1**(1), 104–121 (1998)

In-Car eCall Device for Automatic Accident Detection, Passengers Counting and Alarming

Anna Lupinska-Dubicka[1], Marek Tabedzki[1], Marcin Adamski[1],
Mariusz Rybnik[2(✉)], Miroslaw Omieljanowicz[1], Maciej Szymkowski[1],
Marek Gruszewski[1], Adam Klimowicz[1], Grzegorz Rubin[3], and Khalid Saeed[1]

[1] Faculty of Computer Science, Bialystok University of Technology, Bialystok, Poland
a.lupinska@pb.edu.pl
[2] Institute of Informatics, University of Bialystok, Bialystok, Poland
m.rybnik@ii.uwb.edu.pl
[3] Faculty of Computer Science and Food Science,
Lomza State University of Applied Sciences, Lomza, Poland

Abstract. The European eSafety initiative aims to improve the safety and efficiency of road transport. The main element of eSafety is the pan European eCall project - an in-vehicle system which idea is to inform reliably and automatically about road collisions and even very serious accidents. As estimated by the European Commission, the implemented system will reduce services' response time by 40%. This would probably save 2,500 people a year. In 2015 the European Parliament adopted the legislation that from the end of March 2018 all new cars sold in EU should be equipped with the eCall system. The limitation of this idea is that only a small part of cars driven in UE are sold yearly (about 3.7% cars in 2015). This paper presents the details of concept of an on-board eCall device which can be installed at the owners' request in used vehicles. Proposed system will be able to detect a road accident, indicate the number of vehicle's occupants and send those information to dedicated emergency services via duplex communication channel. This paper presents (1) the basis of the system, (2) the details on accident detection algorithms and hardware used experimentally and (3) state of the art and chosen approach for human detection in vehicle environment.

1 Introduction

In 2016 in the European Union around 25,500 people lost their lives in car accidents [1] whilst in the same time in United States of America, 37,806 people died due to the injuries sustained in vehicle crashes [2]. Diversified studies have shown that there is a possibility to decrease the number of victims by early notification of emergency services [3]. In European Union there is an eCall initiative that brings rapid assistance to people involved in collision. In the case of an accident, this solution will notify emergency services about it. Of course,

© Springer-Verlag GmbH Germany, part of Springer Nature 2020
M. L. Gavrilova et al. (Eds.): Trans. on Comput. Sci. XXXV, LNCS 11960, pp. 36–57, 2020.
https://doi.org/10.1007/978-3-662-61092-3_3

driver can easily turn off notification if he is sure that no one needs help. Due to the European Union directive, from 2018 each new car has to be equipped with eCall emergency system [4]. There is still a problem however: both in EU and US the average age of motor vehicle is about 11 years [5,6]. It means that most of the cars will still not possess such a solution onboard.

In this paper, we would like to present a solution that can be used in relatively old cars. It is an eCall-compliant device that can be installed as an additional unit in diversified motor vehicles. In the proposed approach there is no need to use data from the onboard computer of the car but only sensors permanently mounted in the device itself or additionally installed in the vehicle. In this work, we would like to present algorithms that are useful for collision detection and for passengers detection. This data is valuable also for emergency operator - he will be able to decide which rescue services should intervene at an accident scene. This paper is a large extension of concept presentations in [28] and [29], with detailed explanations of accident simulation and occupants' detection.

This work is organized as follows: in Sect. 2 the authors present principles of eCall initiative with its main goals and requirements. In Sect. 3 the authors describe their preliminary approach. For this aim, the authors explored the existing sensors and algorithms for accident detection (Sect. 4) and occupants detection (Sect. 5), implemented and tested the selected solutions in these areas.

2 eCall System

Studies have found that getting immediate information about an accident and pinpointing the exact location of the crash site can cut emergency services' response time by 50% in rural and 60% in urban areas, leading to 2,500 lives saved per year across the European Union [3]. The eCall system, a pan European emergency notification system, is expected to reduce the number of fatalities in the Union as well as the severity of injuries caused by road accidents, thanks to the early alerting of the emergency services. On 28 April 2015, the European Parliament adopted the legislation on eCall type approval requirements and made it mandatory for all new models of cars to be equipped with eCall technology from 31 March 2018 onward.   .

eCall is an in-vehicle road safety system which idea is to inform about road collision or serious accident [3]. Since eCall is a standard voice call, it can be triggered manually by vehicle driver or one of passengers. However, in case of unconsciousness or inability to move of the vehicle occupants the system is able to automatically contact the nearest Public Safety Answering Point (PSAP), operating within the system of the pan-European 112 emergency network. After establishing the connection eCall system transmits a certain package of basic information about vehicle and accident location (160 bytes). This package, called Minimum Set of Data (MSD) [7], contains among others: latitude and longitude of the vehicle, the triggering mode (automatic or manual) and the Vehicle Identification Number and other information. The purpose is to enable the emergency response teams to quickly locate and provide medical and life-saving assistance to the accident victims. The MSD format has been specified in details in the EN 15722 standard (see Table 1). The MSD package can also contain some additional data previously gathered by specialized sensors installed inside the car that can help the PSAP operator to decide on the emergency services being sent. Such sensors may include Passive Infrared Sensor (PIR) sensors, cameras, seat load sensors, etc.

Table 1. Content and format of Minimal Set of Data MSD (EN15722)

Block no.	Name	Description
1	Format version	MSD format version set to 1 to discriminate from later MSD formats
2	Message Identifier	incremented with every retransmission
3	Control	Automatic or manual activation, position trust indicator, vehicle class
4	Vehicle ID	VIN number according to ISO 3779
5	Fuel type	Gasoline, diesel, etc.
6	Time stamp	Timestamp of incident event
7	Vehicle Location	Position latitude/longitude (ISO 6709)
8	Vehicle direction	2° degrees steps
9	Recent Vehicle Location n-1	Latitude/longitude Data
10	Recent Vehicle Location n-2	Latitude/longitude Data
11	No. of occupants	Minimum known number of fastened seatbelts, omitted if no information is available
12	Optional additional data	e.g. passenger data

3 Authors' Approach

The conclusion that can be drawn from the statistics is clear - the earliest notification of the emergency services of the event determines the survivability of the accident. The solution is eCall system. Not every road user however wants or affords a new car that would be equipped with it. Hence, the authors of the paper have designed a compact system that could be installed in any vehicle. This would allow every car user to rely on the extra security that it provides, for a relatively small price. This chapter presents a general description of the device's operating concept and design. The scheme of the system is depicted in Fig. 1.

　　The first, most important module is the accident detection module. Its task is to detect a dangerous event and launch the entire notification procedure for

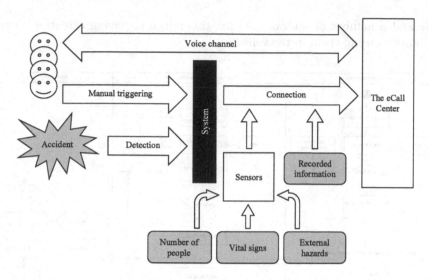

Fig. 1. System block diagram

the relevant service. The key problem here is determining how the accident can be defined and how it can be identified (the devices and parameters used). The system, as requested, should also allow manual triggering. This can be used in the following situations: (1) the accident is not serious enough to trigger the sensors, (2) there is no collision, but occupants gravely need help, (3) the occupants are witnessing an accident and want to inform the eCall services about it.

Another module is the communication module that sends the call to the center. Upon request, additional information recorded in the system (e.g. vehicle data) or read by the sensors (e.g. vehicle location and driving direction read from the GPS receiver) is sent to the center. In addition, the device should be able to establish a voice call with the PSAP operators, allowing them to contact the victims.

The other modules of the system are designed to recognize the situation inside the vehicle to send not only event information to the eCall center, but also assessment of the situation. The first of these modules will detect presence and number of people in the vehicle. The task is actually performed before the event occurs - counting when the occupants occupy or leave the seats - to know how many people were in the vehicle at the time of the accident. In addition, post-event monitoring may be provided to inform eCall operators when occupants have left the vehicle (or, for example, been thrown out if a collision has occurred and they have not fastened their seatbelts). Further work will examine the practicality of such a solution.

A separate, but also important, task will be to identify the vital signs of occupants after an accident. If it is possible to assess whether the occupants are alive or unconscious, it will be valuable information for the services preparing to send help. The authors consider a number of methods for evaluating the state of life and a number of sensors used for this purpose, paying attention to the possibility of using them in the vehicle.

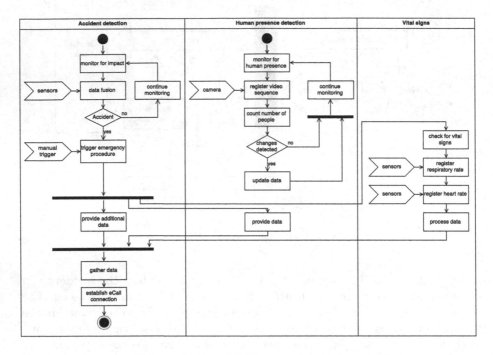

Fig. 2. The activity diagram of the system

In addition to information on the number and severity of injuries, it is possible to collect and share a number of additional information from internal and external sensors - this will be explored and considered in further work. Any details about the condition of the vehicle, the situation inside and also outside (e.g. traffic, external hazards) can be very valuable. All of these additional modules will need to be tested and checked for possibility of use in the intended device - reliability, cost, performance, and ease of installation are the key. This impose restrictions on usable sensors. The proposed system assumes that there is no need to use data from the onboard computer of the car, but only sensors permanently mounted in the device or installed in the vehicle itself.

The project described in this paper assumes yet another kind of information sent to the eCall center - these are all kinds of data input by vehicle users that aim to better and more accurately inform eCall and rescue services. This may include information about illnesses, pregnancy, children's presence, blood type, etc. Providing this information would not be mandatory, but system users would be given the opportunity if they think this could improve their safety. The proposed system will therefore include: a GPS vehicle positioning system, set of sensors for accident detection, a digital infrared camera (or camera system) for detecting vehicle occupants presence and a set of sensors for analyzing passengers' vital functions. In the opinion of the authors, in order to increase the reliability of the system, redundant sensors of the same type should be used in the device. The activity diagram of the system is presented in Fig. 2. The main function of the system will be continuous reading of sensor data and analysis of whether the current values of the measured signals do not exceed the acceptable signal levels. In this case, the system will trigger the automatic accident notification. Simultaneously, the system continually processes the data that will be used to generate the MSD package.

4   Accident Detection

This section presents in details the problem of collision detection: first the state of the art, then the method chosen by the authors and testing environment and finally the obtained results with commentary.

4.1   The State-of-the-art

The problem of collision detection is known in the automotive industry for many years. The need for accident detection systems arose in connection with the idea of using an airbag to protect people in the car. It happened in the 1960s. Despite the passage of nearly 60 years, the issue of detection of an accident is still valid. Although airbag triggering solutions are commonplace every day, publications about new accident detection solutions appear each year. This is due to the development of the concept and practice of Intelligent Transport Systems, the idea of autonomous vehicles, the need to reduce the number of victims in collisions and road accidents, the emergence of new ideas for the use of an accident signal. In the described work, the authors developed a system (hardware and software) for use in a device compatible with the eCall standard and the key assumption was to adapt to work in older models of passenger cars without connecting to the vehicle's electronic systems. During related works analysis, the main focus was on systems which apply the accelerometer as the main sensor used for collision detection. In [8] the device accelerometers with a range of $+/-$ 16 g were used, and these data were limited to the threshold of 3 g, to check whether the proposed algorithm will work correctly in the process of detecting car accidents using a smartphone. Data was taken from the sensors in the phone to construct a dynamic time warping algorithm (DTW) for accident detection. DTW is a time series comparison algorithm originally developed for speech recognition. Two sequences of feature vectors are compared by warping the time axis until an optimal fit (depending on the relevant markers) is found between the two sequences. The formulas were given in the paper and the assessment of the method's effectiveness was shown.

In the accident detection method discussed in [9], the front bumper sensor and the position sensor together with the accelerometer sensor were used to increase the effectiveness of collision detection. The bumper sensor informs the microcontroller what force affects the bumper. Of course, it is significantly greater in the collision event comparing to a parking event. The position sensor is used to determine the occurrence of a drastic change in speed. The information is additionally supplemented with data from the accelerometer (the appearance of sudden acceleration changes).

The paper [10] proposes the use of an IMU (Inertial Measurement Unit) low-cost sensor. An AHRS (Attitude Heading Reference System) sensor was also used to determine the orientation of the vehicle. To analyze the accelerometer, gyroscope and magnetometer data, DCM (Direction Cosine Matrix) was used. After detection of the deceleration above 5 g, the system proposed in [10] checks the vehicle speed. Confirmation of the accident occurs if the speed is less than 5 kph.

The detection method described in [11] uses the ADXL335 module. It is a small, complete, 3-axis accelerometer with voltage outputs. Accelerations in the range of ±3 g are checked. The X axis is monitored and continuously verified whether the value of acceleration changes. In addition, the LM35 temperature sensor is used to continuously monitor the temperature of the surface on which it is attached, e.g. the engine and the bodywork. The infrared sensor module is also used to detect a fire inside the vehicle. The final information is created on the basis of a combination of signals from the above mentioned sensors.

In paper [12], to detect the accident, a 3-axis MMA7660FC sensor is used, which allows detecting various changes such as tilt, tap, shock, etc. The combination of information triggers a signal informing about the occurrence of a collision. In work [13] a tilt sensor was used to detect the accident. The detection method was based on the fact that during a collision a "jump" of the car usually occurs (for a moment some of the wheels lose contact with the surface) or even rollover. The angle of rotation relative to the ground is detected and its value indicates the occurrence of an accident.

Paper [14] concerns the development of an accident detection sensor in situations difficult to clear assessment. The described sensor is adapted to detect frontal and side impact. In addition, the sensor allows to determine the strength and intensity of the collision. An integral part of the structure is a microprocessor with advanced software analyzing and processing in real time all measured parameters and generating a final information of the accident occurrence.

The detection system presented in [15] uses a combination of an accelerometer with a GPS (Global Positioning System). The Kalman filter, the HI-204III GPS receiver and the ADXL345 accelerometer were used. The accident detection scenario is triggered when the vehicle speed exceeds 23 kph. In case of a frontal collision, any negative acceleration greater than 5 g is considered an accident. After detecting a deceleration less than 5 g, the system checks the speed. If the speed drops below 5 kph, the system confirms the accident detection.

In [16], the collision detection module consists of an accelerometer with a large measuring range - MMA621010EG and a relatively small measuring range - MMA7260QT. MMA621010EG is a special car accident detector that is an integrated XY accelerometer. It has a self-test and advanced calibration functionalities. MMA621010EG automatically detects the collision rate (e.g. parking event, serious collision) and rollover, while the MMA7260QT accelerometer detects inclination, changes in orientation in space, shocks and vibrations. An accident is detected on the basis of signals' processing from both sensors.

In road collision detection systems, sensors of mobile devices such as smartphones are also increasingly used. Work [17] presents the WreckWatch system which uses an accelerometer built into the mobile phone. The acceleration filter prevents false calls from being triggered. Filtering alone does not eliminate all false alarms, such as a fall of the phone inside the vehicle or sudden stop. WreckWatch eliminates this type of problems by ignoring all events with deceleration below 4 g. This value is intended to detect even minor accidents but excludes a fall or sudden stop and it was based on empirical analysis. It is also interesting that authors add information known as acoustic events (noise) as an aid to the detection criterion.

The work [18] presents a system called CADANS in which the following sensors were used: accelerometer, GPS - for speed testing, microphone – for noise detection (threshold is 140 dB), camera - for taking pictures or movie. The most important factor used by CADAN to detect car accidents is exceeding the acceleration value above 4 g. Authors claim that the delay threshold of 4 g is not sufficient evidence to conclude that there was a car accident, therefore additional parameters were introduced. Two types of accidents were distinguished: at speeds above and below 24 kph. An interesting idea is to add the "speed variation period" parameter, which measures the value of speed fluctuations in a given time interval.

The analysis of available studies showed that there is no complex systems integrated with the eCall system where not only the accident detection information is sent to emergency services but also the number and life activity of vehicle passengers. In the research process using the hardware platform it is planned to select the target sensor (or group of sensors - in accordance with the Data Fusion approach), ensuring the correct operation of the hardware model.

4.2   Accident Detection Method

There are two methods of accident detection implemented in the proposed system: one for collision detection and second for the rollover detection. The method of collision detection is based on exceeding the overload acceleration threshold. It can be expressed by the following formula, data are provided from the accelerometer readings:

$$M = \sqrt{A_x^2 + A_y^2} \tag{1}$$

where: M - magnitude of acceleration vector, A_x - x-axis acceleration, A_y - y-axis acceleration.

The rollover detection method is based on exceeding the threshold of the vehicle's angle of inclination. It consists in the continuous measurement of the angle of inclination of the vehicle with respect to the vertical in two X and Y axes to detect the rollover. The data comes from the accelerometer and gyroscope, where the ones that derive from the gyroscope are processed with the following formulas:

$$\alpha_{xn} = \alpha_{x(n-1)} + G_x/F \tag{2}$$

$$\alpha_{yn} = \alpha_{y(n-1)} - G_y/F \tag{3}$$

where: α_x - current pitch angle, α_y - current roll angle, $\alpha_{x_{n-1}}$ - previous pitch angle, $\alpha_{y_{n-1}}$ - previous roll angle, G_x - x-axis angular velocity, G_y - y-axis angular velocity, F - frequency of sampling.

However, the data from the accelerometer is converted using the following formulas:

$$\alpha_x = tan^{-1}(a_x/a_y^2 + a_z^2) * 180/\pi \tag{4}$$

$$\alpha_y = tan^{-1}(a_y/a_x^2 + a_z^2) * 180/\pi \tag{5}$$

where: a_x, a_y, a_z - acceleration values.

At the end, for each axis, the above data is combined using the following first order filter:

$$\alpha = k \cdot (\alpha_G \cdot dt) + (1 - k) \cdot \alpha_A \tag{6}$$

where: α_G - angle derived from gyroscope data, α_A - angle derived from accelerometer data.

The main problem of the second method is the gyroscope drift, which generates an error increasing over time. For this purpose, further research should be performed to neutralize or minimize the effect of drift on the algorithm's operation.

Individual steps performed by the data acquisition and sensor handling loop are presented in Fig. 3.

The step to query the sensors is related to the communication with the sensors. If new data occurs, it is retained for the next steps and an appropriate flag is activated to inform about new data.

The next calculation step involves using updated linear and angular acceleration values from the accelerometer and gyroscope. After calculations, when the calculated values are higher than the set thresholds, the procedure appropriate for accident detection is started - putting into queue the information about the accident to the external application and setting the appropriate flags in memory.

The loop attempts to maximize the bandwidth of interfaces used to communicate with sensors and the frequency of measurement updates. During actual measurements, it was possible to obtain data updates at a frequency above 400 Hz using 2 sensors. Obtaining higher frequencies is potentially possible, however, it would require additional work related to verification and possible optimization of the current usage of the communication interface with sensors.

4.3   Testing Environment

Gathering real data from collision detection sensors requires accident environment reproduction. The easiest and cheapest way to verify the accident detection module is to collect real data from collision detection sensors using a suitable

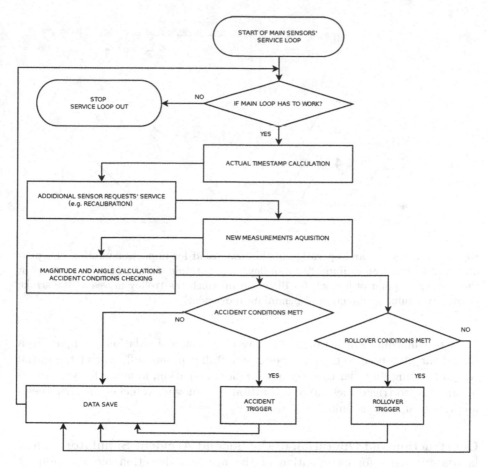

Fig. 3. Steps in the operation of the data acquisition and sensor loops in the program

accident simulator (impact and/or rollover). To this end, laboratory test stands (accident simulators) were built to collect real sensor data, select appropriate sensors, and test algorithms detecting the occurrence of an accident.

Construction and Operation of the First Accident Simulator. The operation of the first impact simulator (Fig. 4) consists in manually pivoting the pendulum to a fixed angular position (max 90°). The pendulum length is about 85 cm. The position of the pendulum can be adjusted, which allows for modification of the impact force in the rear part of the trolley, where the tested electronic system (accelerometer, gyroscope) is placed. Upon its release, the pendulum hits the rear of the trolley. The impacted trolley achieves rapid acceleration and moves along the sliding mechanism. The swing of the pendulum can also be absorbed by means of exchangeable elastic elements of various thickness, stiffness and mechanical characteristics.

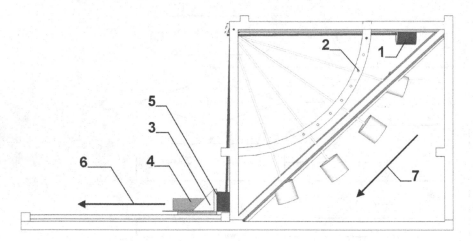

Fig. 4. Construction and operation of the first stand for impact simulation (1 - pendulum, 2 - regulation of impact, 3 - trolley, 4 - electronic circuits (accelerometers) in the trolley, 5 - point of impact, 6 - direction in which the trolley moves at the stroke of the pendulum, 7 - direction of pendulum movement)

A test machine which works this way allows only simulation of impact in a halted car (it is used to trigger accelerations), but it was sufficient at the initial stage of testing. In order to carry out further simulation, a device for simulation of impacts (to trigger delays) and rotation of a moving vehicle - another (new) laboratory stand was built.

Construction and Operation of the Second Accident Simulator. A new laboratory stand for examination of the accident detection module (impact and/or rollover) is shown in Fig. 5. The device is easy to transport, durable, reliable, simple and safe to use. Its dimensions are $182\,cm \times 80\,cm \times 65\,cm$ (length × width × height).

The operation of the second laboratory stand consists of three stages.

1. Manual tensioning of the rubber expander assembly together with the pusher element and immobilizing them in the latch mechanism. The tension force can be changed by increasing or decreasing the number of expanders, by using different types of expanders, and by changing the mounting points.
2. After releasing the latch mechanism, the pusher element starts to accelerate rapidly along the sliding mechanism towards the collision wall. If the trolley with the rotating platform is attached to it before releasing the latch mechanism, the trolley is accelerated together with the tested systems. If the trolley is positioned remotely from the pusher element before releasing the latch mechanism, it strikes the rear of the trolley. The impacted trolley achieves rapid acceleration and moves towards the collision wall. In both cases, the simulation of the accident can be supplemented by the process of automatic rotation of the platform with the tested system - the "rollover"

effect. Turning on the platform rotation enables proper positioning of the structure extending the release bar.

3. The speeding trolley hits the wall. The impact force depends on the tractive force of the drive system and the selected simulation mode (acceleration of the trolley from the beginning, impact of the remote trolley, process of rotating the measurement platform). Stroke impact can also be absorbed by means of exchangeable elastic elements of various thickness, stiffness and mechanical characteristics, which are attached to the collision wall. In this way, the duration of the collision and the occurring overloads can be adjusted.

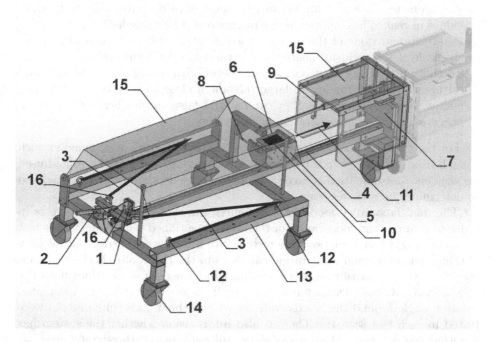

Fig. 5. Construction of a second stand for simulation of collisions and rotations (rollovers) (1 - pusher element, 2 - latch mechanism, 3 - rubber expanders (which generate the driving force of the trolley), 4 - sliding mechanism, 5 - trolley, 6 - swivel platform, 7 - crash wall, 8 - swivel bar releasing the rotating platform, 9 - construction protecting the rotating platform release bar, 10 - electronic circuits in the trolley on the rotating platform, 11 - movement direction of the trolley upon impact, 12 - handles for adjusting the expander tension force, 13 - steel frame, 14 - blocked wheel, 15 - protective cover, 16 - element protecting against accidental release of the drive system)

4.4 Results

For the final version of the device, an accelerometer system for a large measuring range H3LIS331 (up to 200 g) and an IMU LSM6DS33 system containing an accelerometer with a small measuring range (up to 8 g) and a gyroscope were selected. The most important features of the H3LIS331 chip are: dynamically selected measuring ranges (± 100 g/± 200 g/± 400 g), digital I2C/SPI output interfaces and resistance to overload and shocks up to 10000 g. The most important features of the LSM6DS33 chip are: low power consumption for the accelerometer and gyroscope, FIFO buffer up to 8 KB in size, selectable measurement ranges of the accelerometer(± 2/± 4/± 8/± 16 g), selectable gyroscope measuring ranges(± 125/± 250/± 500/± 1000/± 2000 dps), serial interfaces SPI/I2C and built-in temperature sensor. Two accelerometers with different measuring ranges were used, because:

- accelerometer with a small measuring range does not give real acceleration values in real collisions (measurement range will be exceeded),
- the maximum value of the measured acceleration will be used to determine the scale of the accident, hence the need to read the actual values,
- fluctuation readings from the accelerometer with a large range at rest and during normal driving are so large (although close to a percentage of measurement uncertainty) that they can cause a false alarm - hence the need to observe the value of acceleration at low values.

For the above reasons, it is necessary to use an accurate accelerometer with a small measuring range in the case of normal driving, and if the threshold used for accident discrimination is exceeded - also a large accelerometer must be used to determine the accident scale.

The final laboratory model of the device was subjected to further tests in order to examine the cooperation of the above mentioned sensors in various test scenarios. Eight test scenarios were tested, 2 to 6 tests were carried out for each of them. The maximum acceleration values from the H3LIS331 (M1) accelerometer with a large measuring range, the maximum acceleration values from the LSM6DS33 (M2) accelerometer with a small measuring range, the calculated rotation angle around the X axis (roll angle) and the Y axis (pitch angle) were tested in each test scenario. There is also information whether the system has reported the detection of an accident or rollover. As the threshold values, an acceleration of 4.5 g was selected to detect the accident and a roll angle of 60°. The following Tables 2 and 3 show the values of the above-described parameters and whether the system detected a collision (Acc. det. = 1) or detected a rollover (Roll det. = 1).

When analyzing the results of the research, the following conclusions can be drawn:

- in all cases examined, the collision was detected,
- in all cases of rollover simulation the rotation angle (roll) was greater than 60° and rollover was detected,
- the resultant value of the acceleration depends on the initial speed (the number of expander used)
- distractions caused by drift or small changes (vibrations) are noticed, especially for the H3LIS331 accelerometer,

Table 2. Experimental results from scenarios without rollover

Scenario	Test no.	M1	M2	Roll angle	Pitch angle	Acc. det.	Roll. det.
5 ropes no roll	153208	20,63	8,20	15,46	3,40	1	0
	153156	27,16	8,08	13,05	3,12	1	0
	153208	20,63	8,20	15,46	3,40	1	0
	153217	24,47	8,23	9,37	2,32	1	0
	153226	28,28	8,32	12,88	2,87	1	0
	Mean	**24,23**	**8,21**	**13,24**	**3,02**		
9 ropes no roll	152830	48,35	8,10	22,34	4,86	1	0
	152913	47,02	8,09	27,27	6,19	1	0
	152936	49,72	9,13	14,87	3,28	1	0
	152953	33,11	8,08	15,66	3,15	1	0
	153006	32,27	8,32	16,32	3,65	1	0
	Mean	**42,10**	**8,35**	**19,29**	**4,23**		
5 ropes rear hit then front hit no roll	153818	22,77	8,00	12,96	3,20	1	0
	153838	17,99	8,12	13,66	3,86	1	0
	154049	22,95	8,15	9,76	2,84	1	0
	154104	19,00	8,45	8,77	3,64	1	0
	154111	24,78	8,01	10,59	1,37	1	0
	Mean	**21,50**	**8,14**	**11,15**	**2,98**		
9 ropes rear hit then front hit no roll	154855	23,08	9,96	9,87	3,00	1	0
	154941	31,66	8,11	11,77	4,81	1	0
	154953	32,24	9,51	8,33	8,38	1	0
	155005	30,89	11,31	8,65	15,97	1	0
	155022	30,06	11,31	12,39	24,10	1	0
	155051	31,86	10,68	17,10	12,77	1	0
	Mean	**29,96**	**10,15**	**11,35**	**11,50**		

– it is suggested to carry out an accurate calibration and possible elimination of minor shocks when measuring the acceleration,
– the gyroscope correctly determines the angle of rotation,
– the calculated rotation angle is also disturbed by drift - it can be seen for collisions without rotation where the maximum angle in a few seconds increases to several degrees,
– it is suggested to conduct an accurate calibration of the gyroscope,
– it would also be necessary to check the dependence of the sensors' indications on the temperature and introduce some corrections - especially when it regards to the zero-scaling.

Table 3. Experimental results from scenarios with rollover

Scenario	Test no.	M1	M2	Roll angle	Pitch angle	Acc. det.	Roll. det.
5 ropes with roll before hit	153346	26,48	10,67	224,61	4,29	1	1
	153432	22,35	9,50	230,55	3,88	1	1
	153456	21,98	8,55	237,21	4,32	1	1
	153521	29,98	11,07	234,00	3,58	1	1
	153544	28,62	8,41	236,63	4,05	1	1
	Mean	**25,88**	**9,64**	**232,60**	**4,03**		
9 ropes with roll after hit	152440	31,24	8,18	206,80	12,64	1	1
	152551	33,90	9,26	196,78	5,55	1	1
	Mean	**32,57**	**8,72**	**201,79**	**9,10**		
5 ropes rear hit then front hit with roll	154153	24,50	10,76	224,48	2,80	1	1
	154213	22,95	8,65	256,01	3,53	1	1
	154325	21,19	10,13	252,35	4,22	1	1
	154433	19,45	10,10	224,65	2,11	1	1
	154447	21,63	8,07	236,63	7,33	1	1
	154515	23,70	10,09	231,25	3,71	1	1
	Mean	**22,24**	**9,63**	**237,56**	**3,95**		
9 ropes rear hit then front hit with roll	154617	23,94	8,00	260,65	5,91	1	1
	154641	31,65	11,31	248,45	3,28	1	1
	154706	28,94	11,31	239,53	5,18	1	1
	154724	29,51	9,26	240,42	3,20	1	1
	154759	30,66	8,04	258,42	6,90	1	1
	Mean	**28,94**	**9,58**	**249,49**	**4,89**		

Table 4 shows the minimum and maximum acceleration values and the magnitude calculated according to Eq. (1) when driving with a real vehicle with a detection system installed on board. Twelve series of experiments were carried out. Each series corresponded to the driving in different road conditions and at different speeds: stop, slow passage through residential streets (30 kph), normal city driving (50 kph), driving out of town at a higher speed (up to 90 kph), driving a car on an uneven surface and emergency braking (from 60 kph to 0).

Analysis of Table 4 shows that for none of the test cases during the test drive, the threshold value of 4.5 g has not been reached, even in the event of sudden braking and driving on uneven surfaces, so **the false positives rate is 0%.**

Table 4. Acceleration values obtained from the LSM6DS33 sensor while driving in different conditions

Series	X max	X min	Y max	Y min	Z max	Z min	Magnitude
1	0.144	−0.144	0.144	−0.144	1.152	0.864	0.161
2	0.504	−0.432	0.648	−0.648	1.800	0.360	0.709
3	0.576	−0.504	0.576	−0.504	1.440	0.576	0.664
4	0.504	−0.504	0.792	−0.720	1.728	0.288	0.805
5	0.432	−0.360	0.504	−0.504	1.656	0.504	0.519
6	0.576	−0.648	0.576	−1.224	2.232	0.000	1.224
7	1.152	−0.144	0.864	−1.008	1.368	0.504	1.207
8	1.152	−0.144	1.080	−1.080	1.368	0.576	1.163
9	0.936	−0.144	1.152	−1.152	1.512	0.288	1.207
10	0.648	−0.648	0.864	−1.080	2.160	−0.144	1.163
11	1.224	−0.504	0.792	−1.008	2.016	0.000	1.239
12	0.864	−0.432	1.800	−0.720	1.800	0.144	1.806

5   Human Presence Detection

As a part of the proposed in-vehicle system, it is required to detect the number of occupants and possibly their vital condition. In this subsection the literature review of the solutions connected with human presence detection and our idea how to solve this problem are presented.

There are many different approaches to human detection problem but small number of them is associated with the vehicles. An interesting idea was presented in [19]. The authors of this solution proposed the system that detects moving objects in a complex background. In this paper Gaussian Mixture Model (GMM) was claimed as an effective way to extract moving objects from a video sequence. The authors also claimed that conventional usage of GMM suffers from false motion detection in complex background and slow convergence. To achieve robust and accurate extraction of moving objects a hybrid method by which noise is removed from images was used. Proposed model first stage consist of fourth order Partial Differential Equation (PDE). As the second stage usage of relaxed median filter on previously obtained output is claimed. The authors provided experiments results in which it is clearly visible that their approach has detection rate equal to 98,61% what compared to conventional usage of GMM is more convenient, forasmuch in this paper GMM was claimed to has 73,5% accuracy.

The authors of [20] claimed that technology development enabled usage of human detection methods in the intelligent transportation system in smart environment. In this paper it was pointed that it is a huge challenge to implement an algorithm that will be robust, fast and could be used in Internet of Things (IoT) environment. In the case of these systems, low computational complexity is crucial due to available resources. The authors proposed to use Histogram of Oriented Gradients (HOG) method. Conventional usage of this algorithm gives accuracy on satisfactory level although it is really expensive to compute. The described solution aims to reduce the computations using approximation methods and adapts for varying scale. In this work it was pointed that experiments were done on online available datasets like the one that is shared by Caltech. This approach uses also specific hardware solutions and it is based on FPGA structures. In the paper experiments results were presented. In the case of database consists of about 1000 samples, accuracy level was equal to 95,57% when the database has grown to 2000 samples, accuracy level has decreased to 86,43%.

Both of recently proposed solutions are not connected with the vehicles but it is considered that it is possible to adjust them to use in the car environment. Another concept that the authors would like to present is the one described in [21]. This article provided an idea to create the system for people counting in public transportation vehicles. This concept was created due to the problem of monitoring the number of occupants getting in or out of public transportation in order to improve vehicle's door control. In this work both hardware and software implementations are combined. This approach uses stereo video system that acquires a pair of images. In the next steps Disparity Image computation and 3D filtering are claimed to be used. In this article no experiments results were provided however it is claimed that the system was installed and tested in natural environment in the bus.

An interesting approach was also described in [22]. In this case people were not counted but the authors take into consideration automatic traffic counting. Aim of the experiments was to define the requirements to be applied to traffic technologies to match specific applications. The authors tested seven different traffic monitoring systems. In the paper it is claimed that the study lasted one year. As a result the specific test procedures were developed and the most reliable vehicles counting system was chosen. The authors of [22] inform that vehicles were counted manually and then the result was compared with automatic analysis of the recorded video.

Recent advances in deep neural networks allow to perform object detection in images with a level of accuracy comparable to human performance. In our application we investigated object detection techniques based on deep neural networks to count the number of persons in a car using images from camera installed in a vehicle. We investigated two approaches for person counting: detection of faces and detection figures. Initial experiments were conducted using pretrained models that are available in OpenCV [23] package. For a face detection we used the Single Shot MultiBox Detection (SSD) algorithm [24]. Detection of persons' figures was performed with You Look Only Once (YOLO) method [25].

Fig. 6. Face detection using SSD method in an image from infrared light wide angle camera (Waveshare HD Night Vision E OV5647) with LED illumination. Detected face location is marked with red box. (Color figure online)

During experiments various camera types were tested. The visible light cameras are most widespread and least expensive devices. Their main drawback is poor performance in low lighting conditions. One may solve this problem by using infrared cameras. They allow to use infrared illumination that is not visible by humans and therefore does not distract the vehicle driver and its passengers. The third option is thermal imaging. Such cameras allow to register infrared thermal radiation emitted by warm objects. They also provide information on the temperature which may be utilized to discriminate animate objects from inanimate things, for example: a face of alive human from a face in a photograph. However, thermal cameras are very expensive and their resolution is much lower compared to devices that register visible and reflected infrared light. Another aspect that must be considered is correct exposition. Counting people based on face detection requires that a face front or profile must be visible. In case of human silhouette detection the person body should not be significantly obscured. To ensure sufficient exposition different locations and camera optics were considered during investigation. For a solution based on a single camera two settings were proposed: wide angle camera installed under a rear mirror and 360° camera attached to a roof ceiling between front and back rows of seats.

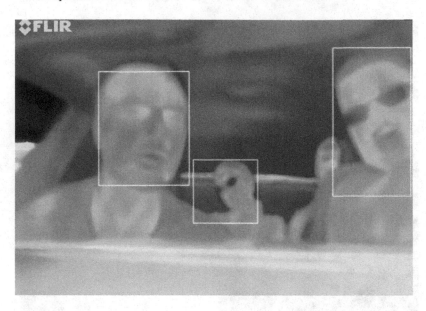

Fig. 7. Face detection using SSD method in an image from thermal camera (Raspberry Pi NoIR Camera HD v2). Detected faces locations are marked with red boxes. (Color figure online)

Figure 6 shows an example of face detection using image from infrared camera installed near rear mirror. In such a position frontal seats are usually well exposed, however faces of passengers sitting in the back may be obscured. Figure 7 shows an example of face detection using image from thermal camera. Obviously thermal camera seems suitable for low-light conditions, however we experienced problems in very hot weather, when the temperature of car components were quite close to human body. Thermal cameras are also rather expensive. Figure 8 shows an example of YOLO silhouette detection using 360° camera (preprocessed from radial to standard coordinates with slight overlapping, also with histogram equalisation). The results are accurate despite various angle expositions and reach 85%.

Summary. The experiments conduced revealed that with adequate light and correct exposition of either silhouettes of faces modern algorithms detect people almost perfectly. Therefore the problem is rather related to people visibility (light, angle, obscurity) than algorithms efficiency. In recommended testing environment (360° camera placed in the center of vehicle with adequate light, YOLO silhouettes detection) the accuracy of passengers detection was close to 85%.

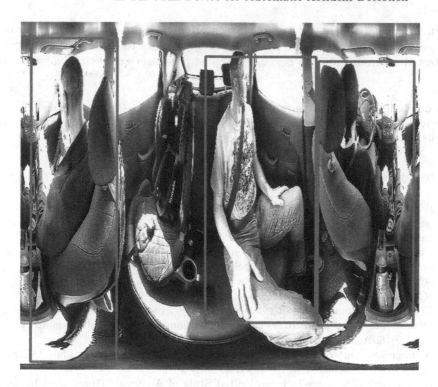

Fig. 8. Person detection using YOLO method in an image from 360° camera (SJCAM SJ360). Detected figures locations are marked with red boxes. (Color figure online)

6   Conclusions and Future Works

eCall seems to be an excellent initiative to save lives, however one must note the limitation of the directive, as new cars sold in EU in 2015 constitute only about 13.7 million versus total of over 270 million cars driven in EU [26,27] (partially estimated). Thus the idea of compact and cheap device adjoined to existing cars is surely a very interesting proposal. In this article the authors have described the eCall solution, mentioned and evaluated algorithms and devices possibly used for such compact eCall-compliant device and presented the ideas and experiments of their approach in subtasks of car accidents detection and passengers detection.

At the moment, the work on the described system is carried out in parallel on two levels: accident detection and analysis of information about vehicle occupants. Two laboratory stations to carry out collision tests have also been developed and are described in the paper. The accomplished stages of accident detection include obtaining data on acceleration/delays from the crash test, selection of sensors (accelerometer and gyroscope to assist in determining the scale of an accident) and establishing thresholds which trigger emergency procedure. Simulated accidents were detected with 100% accuracy. Works on passengers detection required collecting visual data and interpreting it. The authors tested various cameras and expositions, along with DNN approaches to count the passengers. In tested environments the best accuracy of passengers detection was close to 85%.

Acknowledgments. This work was supported by grant S/WI/3/2018 and S/WI/ 2/2018 from Bialystok University of Technology and funded with resources for research by the Ministry of Science and Higher Education in Poland.

References

1. Road Safety: Encouraging results in 2016 call for continued efforts to save lives on EU roads. http://europa.eu/rapid/press-release_IP-17-674_en.htm. Accessed 24 Dec 2017
2. Number of fatal car accidents in United States of America. https://crashstats. nhtsa.dot.gov/Api/Public/ViewPublication/812580. Accessed 15 Sept 2019
3. eCall: Time saved = lives saved. https://ec.europa.eu/digital-single-market/en/ eCall-time-saved-lives-saved. Accessed 15 Sept 2019
4. European Parliament makes eCall mandatory from 2018. http://www.etsi.org/ news-events/news/960-2015-05-european-parliament-makes-ecall-mandatory-from-2018. Accessed 24 Dec 2018
5. Average age of motor vehicles in European Union. https://www.acea.be/statistics/ article/average-vehicle-age. Accessed 25 May 2019
6. Average age of motor vehicles in United State of America. https://www.bts.gov/ content/average-age-automobiles-and-trucks-operation-united-states. Accessed 25 May 2019
7. ETSI eCall Test Descriptions - ETSI Portal. https://portal.etsi.org/cti/ downloads/TestSpecifications/eCall_TestDescriptionsv1_0.pdf. Accessed 24 Dec 2017
8. Aloul, F., Zualkernan, I., Abu-Salma, R., Al-Ali, H., Al-Merri, M.: iBump: smartphone application to detect car accidents. In: IAICT 2014, Bali, 28–30 August 2014
9. Kushwaha, V.S., et al.: Car accident detection system using GPS and GSM. Int. J. Eng. Res. Gen. Sci. **3**(3) (2015)
10. Amin, S., et al.: Low cost GPS/IMU integrated accident detection and location system. Indian J. Sci. Technol. **9**(10) (2016)
11. Sulochana, B., Sarath, B.A., Babu, M.: Monitoring and detecting vehicle based on accelerometer and MEMS using GSM and GPS technologies. Int. J. Comput. Sci. Trends Technol. (IJCST) **2**(4) (2014)
12. Reddy, M.R., Tulasi, J.: Accident detection depending on the vehicle position and vehicle theft tracking, reporting systems. Int. J. Sci. Eng. Technol. Res. **3**(9), 2359–2362 (2014)
13. Islam, M., et al.: Internet of car: accident sensing, indication and safety with alert system. Am. J. Eng. Res. (AJER) **2**(10), 92–99 (2013). e-ISSN 2320-0847, p-ISSN 2320-0936
14. Rich, D., Kosiak, W., Manlove, G., Potti, S.V., Schwarz, D.: A sensor for crash severity and side impact detection. In: Ricken, D.E., Gessner, W. (eds.) Advanced Microsystems for Automotive Applications 98, pp. 1–17. Springer, Heidelberg (1998). https://doi.org/10.1007/978-3-662-39696-4_1

15. Amin, S., et al.: Kalman filtered GPS accelerometer based accident detection and location system: a low-cost approach. Curr. Sci. **106**(11) (2014)
16. Vidya Lakshmi, C., Balakrishnan, J.R.: Automatic accident detection via embedded GSM message interface with sensor technology. Int. J. Sci. Res. Publ. **2**(4) (2012)
17. White, J., Thompson, C., Turner, H., Dougherty, B., Schmidt, D.C.: Wreckwatch: automatic traffic accident detection and notification with smartphones. Mob. Netw. Appl. **16**(3), 285–303 (2011)
18. Ali, H.M., Alwan, Z.S.: Car accident detection and notification system using smartphone. IJCSMC **4**(4), 620–635 (2015)
19. Fazli, S., Pour, H.M., Bouzari, H.: A robust hybrid movement detection method in dynamic background. In: Proceedings of 6th Conference Telecommunications & Information Technology 2009, ECTI-CON 2009, Pattaya, Chonburi, Thailand (2009)
20. Sageetha, D., Deepa, P.: Efficient scale invariant human detection using histogram of oriented gradients for IoT services. In: 2017 30th International Conference on VLSI Design and 2017, 16th International Conference on Embedded Systems Proceedings (2017)
21. Bernini, N., Bombini, L., Buzzoni, M., Cerri, P., Grisleri, P.: An embedded system for counting passengers in public transportation vehicles. In: 2014 IEEE/ASME 10th International Conference on Mechatronic and Embedded Systems and Applications Proceedings (2014)
22. Bellucci, P., Cipriani, E.: Data accuracy on automatic traffic counting: the SMART project results. Eur. Transp. Res. Rev. **2**(4), 175–187 (2010)
23. Vanhamel, I., Sahli, H., Pratikakis, I.: Automatic wathershed segmentation of color images. In: Goutsias, J., Vincent, L., Bloomberg, D.S. (eds.) Mathematical Morphology and its Applications to Image and Signal Processing. Computational Imaging and Vision, vol. 18, pp. 207–214. Springer, Boston (2000). https://doi.org/10.1007/0-306-47025-X_23
24. Liu, W., et al.: SSD: single shot multibox detector. In: Leibe, B., Matas, J., Sebe, N., Welling, M. (eds.) ECCV 2016. LNCS, vol. 9905, pp. 21–37. Springer, Cham (2016). https://doi.org/10.1007/978-3-319-46448-0_2
25. Redmon, J., Divvala, S.K., Girshick, R.B., Farhadi, A.: You only look once: unified, real-time object detection. In: CVPR 2016, pp. 779–788 (2016)
26. https://www.best-selling-cars.com/europe/2016-full-year-europe-best-selling-car-manufacturers-brands . Accessed 18 Nov 2017
27. Eurostat - Passenger cars in the EU. http://ec.europa.eu/eurostat/statistics-explained/index.php/Passenger_cars_in_the_EU. Accessed 18 Oct 2017
28. Szymkowski, M., et al.: The concept of in-vehicle system for human presence and their vital signs detection. In: 5th International Doctoral Symposium on Applied Computation and Security Systems: ACSS2018 (2018)
29. Lupinska-Dubicka, A., et al.: The conceptual approach of system for automatic vehicle accident detection and searching for life signs of casualties. In: Chaki, R., Cortesi, A., Saeed, K., Chaki, N. (eds.) Advanced Computing and Systems for Security. AISC, vol. 883, pp. 75–91. Springer, Singapore (2019). https://doi.org/10.1007/978-981-13-3702-4_5

Volumetric Density of Triangulated Range Images for Face Recognition

Koushik Dutta[(⊠)] [ⓘ], Debotosh Bhattacharjee, and Mita Nasipuri

Department of Computer Science and Engineering, Jadavpur University,
Kolkata 700032, West Bengal, India
koushik.it.22@gmail.com, debotoshb@hotmail.com,
mitanasipuri@gmail.com

Abstract. In this paper, a volumetric space representation of 3D range face image has been established for developing a robust 3D face recognition system. A volumetric space has been created on some distinct triangular regions of the 3D range face image. Further, we have constructed 3D voxels corresponding to those regions for developing voxelization-based 3D face classification system. The proposed 3D face recognition system has mainly three parts. At first, seven significant landmarks are detected on the face. Secondly, any three individual landmarks are used to create a triangular region; in this way, six distinct triangular areas have been generated, where the nose tip is a common landmark to all the triangles. Next, assume a plane at the nose tip level for representing the volumetric space. The total density volume and some statistical features are considered for the experiment. From the volumetric space, construct 3D voxel representation.

Further, geometrical features from 3D voxel are used for the experiment. Three popular 3D face databases: Frav3D, Bosphorus, and Gavabdb are used as the input of the system. On these databases, the system acquires 94.28%, 95.3%, and 90.83% recognition rates using kNN and 95.59%, 96.37%, and 92.51% recognition rates using SVM classifier. Using geometric features with SVM classifier, the system acquires 92.09%, 93.67%, and 89.7% recognition rates.

Keywords: Volumetric space · 3D range face image · 3D voxels · Landmark · Total density volume · Statistical feature · Geometrical features

1 Introduction

Face recognition [1] is one of the most popular biometric research areas having applications in various domains like ATM counter, Airport, Military check post, etc. 3D face recognition overcomes the problems of its counterpart in 2D such as pose and illumination variations. Generally, three different approaches are used for face recognition: holistic, based on local features and hybrid. The holistic methods like Principal Component Analysis (PCA) [2, 3], Linear Discriminant Analysis (LDA) [4], and Independent Component Analysis (ICA) [5], etc. There exists a survey on the local feature-based method for 3D face recognition [6]. Based on different feature-based ways, the authors divided the methods into two parts: local and global. The survey also

© Springer-Verlag GmbH Germany, part of Springer Nature 2020
M. L. Gavrilova et al. (Eds.): Trans. on Comput. Sci. XXXV, LNCS 11960, pp. 58–84, 2020.
https://doi.org/10.1007/978-3-662-61092-3_4

presents state-of-the-art for 3D face recognition using local features. Global features are part of holistic-based approaches. The hybrid-based methods are deified based on the combination of local and holistic-based approaches.

To develop a system, one of the main modules is feature extraction, on which the performance of the system depends. Feature extraction from images is very much challenging task to the researchers. It is challenging to extract innovative features that recognize or verify the person appropriately. On the other side, the accuracy of face recognition depends on the number of features. Our main objective is to minimize the number of features that represent a face reliably. Here, we transform the range face images into a different form or in a separate space, i.e., volumetric space. The alternative representation is essentially capturing the surface information.

The volumetric space defined by calculating the volume of the selected triangular regions on range face images. The present 3D face recognition system generates features from the volumetric space. A preliminary version of this work has been reported [7]. The proposed method is divided into some significant steps. The steps are given below:

1. Register the non-frontal face into frontal position using well-known Iterative closest point (ICP) technique.
2. Seven landmarks are identified on the significant portions like the nose, eye, and mouth region of the input range face. Initially, the pronasal is chosen as the position, where both curvedness and mean curvature are maximum according to depth positions. Next, other landmarks, like eye and mouth corners, are detected using the geometrical calculation of the face, followed by a mask representation for better localization. Pronasal is the point from where geometrical calculations are initiated.
3. Six distinct regions have been created by using any three landmarks, where the pronasal is one vertex of all the triangles considered for feature extraction.
4. The range image is transformed into volumetric form by assuming a plane at the nose tip level.
5. Calculate the total volume of each triangular regions for further classification.
6. For classification, total volume or density of each region is included in the final feature vector. Side by side, we have considered some statistical features like mean, variance, standard deviation, skewness, and entropy of the triangular volumetric space.
7. Construct 3D voxel from the volumetric space to represent the classification system from different view.
8. Introduced some geometric voxel features such as surface area, area normal etc.

The system has some significant contributions:

1. Landmarks detection using the combination of geometrical and 3D shape descriptor-based approach.
2. Creation of triangular regions using three individual landmarks, where the nose tip is the common point in all the triangles.
3. Volumetric space creation of each triangle from the input range image.
4. 3D voxels representation from volumetric spaces of those selected triangular regions.

Various works have already been done on 3D face recognition using range image, or 3D point clouds. Other than the range images, the 3D point cloud is used to create 3D voxel and polygonal mesh representation. Here, the description of a 3D range face image in volumetric space is a new idea for recognition. Previously, there exists another work [8], where the authors have represented the 3D range image in complement component space. The components are calculated along positive and negative X and Y axes.

Further for classification, discriminative attributes have been generated using singular value decomposition (SVD). The experiment gives more detail information from range images. The investigation extracted the components from all types of image include pose, expression, and occlusion. Extraction of components from the occluded face may lead to low recognition performance.

In [9], the authors have represented a 3D voxel of the whole face in three different ways: (1) those consisting of a single cut of the cube; (2) those consisting of a combination of individual cuts of the cube; (3) depth maps. They developed a 3D face recognition using voxel representation with different single cut such as horizontal cut at the level of the nose, eye, and mouth, similarly, vertical cut at the level of nose and eye. They have also used depth maps of full face and half face based on horizontal and vertical. Two separate classifiers, support vector machine (SVM) and Euclidean distance-based k-nearest neighbor (kNN) classifier, were used for classification of PCA-based feature vectors. The authors measured the efficiency of the system using both the 3D voxel and depth map representation. The experiment was tested using the pose and expression variant face images that are acquired in both controlled and uncontrolled environment.

Next, in [10], the authors presented a descriptor, named as, speed up local descriptors (SULD) from significant points extracted for face range image. The Hessian-based detectors were used for detecting the significant points. The SULD descriptor, against all the significant points, is computed for extraction of features. The experiment worked on frontal, and non-frontal images include neutral and expression variant.

In [11], the authors have introduced a system where three different regions: eye, nose, and mouth separately classify the image. Initially, the whole range face image is transformed into its Local Binary Pattern (LBP) form. Further, Histogram of Oriented Gradient (HOG) features are extracted. In that method, occlusion and expression invariant images were considered as input. The authors had separated due to highlighted of any particular region that gives best recognition accuracy, instead of, regarded as whole face region. This system is useful when we don't need a full face for recognition.

In our method also, we have generated 3D voxel from 3D volume space. The voxel representation can be defined as the transformation of a 3D mesh into a regular representation of a fixed number of volume elements.

The rest of the paper is organized as follows: Sect. 2 gives the details description of different steps of the proposed system. Next, the experimental result of our proposed work discusses in Sect. 3. After that, Sect. 4 illustrates the results and comparison with previous techniques. Section 5 concludes the paper.

2   Proposed System

The present proposed system is divided into five significant stages. The flow diagram in Fig. 1 illustrates the individual stages of the system.

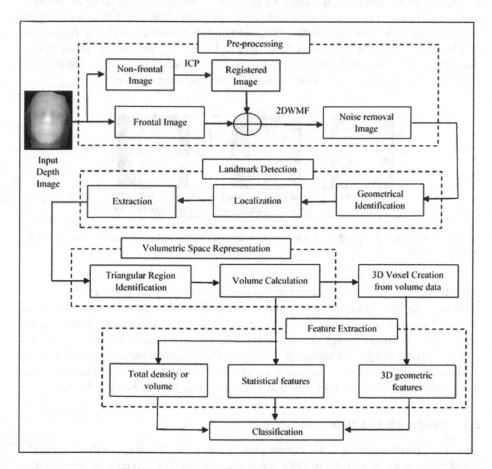

Fig. 1. Flow diagram of the proposed system

2.1   3D Range Image Acquisition

Range or 2.5D or Depth images [12] are the 2D matrix representation of 3D depth values. It is represented as a gray-scale image, where the value is in between 0 and 255 (both inclusive). Various 3D scanners such as structured light, laser scanner, and Kinect 3D scanner are used to capture 3D point clouds. Further, the range image or depth image is generated from a 3D point cloud. The proposed work is using the range image data of three well-known 3D face datasets: Frav3D, Bosphorus, and Gavab.

2.2 Pre-processing

This stage divided into two sub-stages: registration and smoothing. The details of the sub-stages are given below.

Registration of Non-frontal Faces. In this work, we have considered all possible variations, namely pose, expression, and occlusion in the input depth images. For pose variant images, we have used Iterative Closest Point (ICP) technique [14] to register the non-frontal face image to frontal face image. Figure 2 given below illustrates an example of 3D face registration using ICP. The ICP technique tries to find out the rigid transformation that minimizes least square distance between two points.

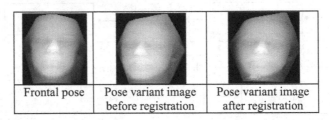

| Frontal pose | Pose variant image before registration | Pose variant image after registration |

Fig. 2. Non-frontal 3D depth face image before and after registration. Image is taken from the Frav3D dataset

Smoothing. At the time of scanning by the 3D scanner, the point cloud contains various noises like spike and holes. There exist different 2D filters like mean, median, weighted median, and Gaussian smoothing, etc., those can be used for removing the spike from the range image. In this work, the weighted median filter is used for removing spike noise from input range images. The weighted median filter [13] is the median filter in a particular case. The mask consists of a wide variety of properties than the median filter. The mask of filter has a non-negative integer weighted value.

2.3 Landmark Detection

Landmark identification is a significant part of our work. Further, we have created a triangular region by using any three landmarks, where the pronasal is the common point for all the triangle. The landmark is 2D or 3D Cartesian coordinate point that has some biological meaning. Soft tissue and hard tissue are responsible for two types of available landmarks. Soft tissue landmarks are identified on human skin, whereas hard tissue landmarks are determined based on the inside structure of the body. Here, we have detected some significant soft tissue landmarks around the nose, mouth, and eye portions on the human face. Landmarks [15, 16] are detected in various ways. Some of the holistic-based approaches are like active appearance model-based, learning-based fitting, and analytic fitting methods do exist. Other than these, local appearance-based methods, as well as regression-based methods, are also used for landmark detection. Currently, deep learning-based techniques are also used for landmark detection.

Here, seven distinct landmarks are detected such as pronasal (nose tip), left and right endocanthions (inner eye corner), left and right exocanthions (outer eye corner), and left and right chelions (mouth corner). The landmarks are extracted from an area covering a significant portion of the range face image. An approach that combines geometrical and 3D descriptor-based approach is used to detect those landmarks. The human facial geometric structure is mainly used to identify eye and mouth corner concerning nose tip, as in [16], where it is shown that the mouth and eye corners are approximately in fixed length with the fixed angle from the nose tip point. Here, the fixed length is denoted as the radius of the circle, say R, around the nose-tip point (a, b) as the center. Equation 1 shows the geometrical formulae for the detection of all other landmarks against the nose tip.

$$x = R\cos\theta + a; \quad y = R\sin\theta + b \quad where \quad \theta = (pi/180) * \emptyset \tag{1}$$

where, $\emptyset$ is the angle between the nose tip and another landmark point.

Using the geometrical approach, the detected eye and mouth corners from nose tip are not always accurate. So, to localize the points more accurately, we have used the 3D shape descriptors [16] such as coefficients of fundamental forms, mean (H), Gaussian curvature (K), maximum and minimum principal curvatures (k_1 and k_2), shape index (SI), and curvedness index (CI). The first and second fundamental forms are defined in Eqs. 2 and 3.

$$Edu^2 + 2Fdudv + Gdv^2 \tag{2}$$

$$edu^2 + 2fdudv + gdv^2 \tag{3}$$

Where E, F, G, e, f, g are the coefficients. All the descriptors [16] are calculated from first and second order derivatives of the input depth image matrix. From all the descriptors, SI is most famous for initial localization after landmarks detection by geometrical approach. As in [16], the range of SI belongs to $[-1, 1)$, which is divided into seven distinct representation, as shown in Fig. 3. A mask of size 7×7 is considered around the detected landmark point for appropriate localization. We have examined all the surface descriptors and derivative values on that mask.

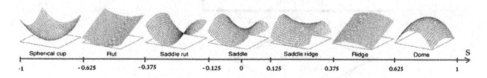

Fig. 3. Seven distinct shape index range

Further, extracted the accurate position of landmark points by taking maximization or minimization of the descriptors within the mask. A block diagram illustrating the proposed approach is given in Fig. 4. After that, the detection technique of each landmark has been discussed.

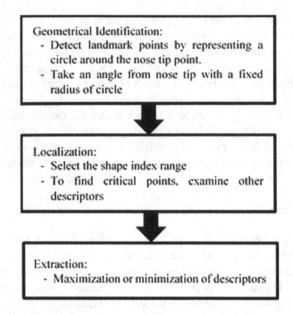

Fig. 4. Block diagram of proposed landmark detection approach

Pronasal. Pronasal is one of the most important facial landmark points on the nose tip, which easily detected by the human eye. In the case of the frontal position of the human face, the point is the shortest distance from the capturing device compared to other location of the human face. In frontal 3D depth face images, this point holds maximum depth value. In some cases, maximum depth value occurs multiple times in the neighborhood of the point. In that case, we have considered a new approach that has calculated the curvedness index and means curvature on those pixels. The point having both curvedness index and mean curvature as a maximum is considered as the pronasal point. The flow diagram is given in Fig. 5. Next, only the nose tip localization is shown in Fig. 6.

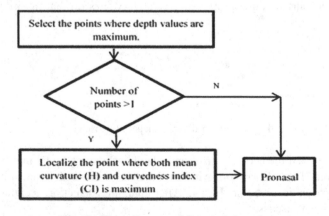

Fig. 5. Flow diagram of Pronasal detection

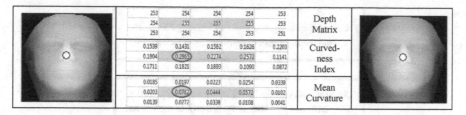

Depth Matrix					
253	254	254	254	253	Depth
254	255	255	255	253	Matrix
253	254	254	253	251	
0.1539	0.1431	0.1562	0.1626	0.2203	Curved-
0.1804	0.2865	0.2274	0.2572	0.1141	ness
0.1711	0.1821	0.1893	0.1090	0.0872	Index
0.0185	0.0197	0.0223	0.0254	0.0339	Mean
0.0203	0.0782	0.0444	0.0572	0.0102	Curvature
0.0139	0.0772	0.0338	0.0108	0.0041	

Fig. 6. Nose tip localization

Chelion. Chelions are the landmark locations on the mouth corner. According to [17], initially, the Chelion points are detected using the geometrical approach. A circle is generated around the pronasal point of radius r is in between 15 to 25. Further, we have considered two separate angles concerning X-axis. For detection of the left mouth corner, considered the angle θ_1 in between 220 to 240. Similarly, for right mouth corner, considered the angle θ_2 in between 300 to 320, as shown in Fig. 10(a) and (b). Further, we have used shape descriptors for appropriate localization. As already said that we have taken 7×7 mask around the detected point for further processing. Consider saddle range value within the mask for the mouth corner. Further, for more localization, reduce the points on the mask by using other descriptors. The first order derivative gives the slope of the curve of the function $f(x, y)$. The first order derivative with respect to x is denoted by f_x. Considered the points where $f_x > 0$ for left mouth corner and $f_x < 0$ for right mouth corner. It reduces some points in the mask. Finally, for the extraction of a landmark point, take a minimum of coefficient g of the second fundamental form to compare to other descriptors. The flow diagram of the chelion detection process is given in Fig. 7.

Endocanthions. Endocanthions are the points where the upper and lower inner eyelids meet at inner eye corners. According to [17], similarly, the endocanthion points are detected using the geometric approach. As per previous, considered a circle of radius r is in between 10 to 15 and two separate angles concerning X-axis. For detection of the left eye inner corner, considered the angle θ_1 in between 110 to 130 and for right mouth corner, considered the angle θ_2 in between 50 to 70 as shown in Fig. 10(c) and (d). Further, we have used shape descriptors for appropriate localization. Here, the SI belongs to the ridge surface [0.375, 0.625). Further, the first order derivative with respect to x is greater than zero for both left and right inner eye corner. For more optimum localization, considered the coefficients of first and second fundamental form $F > 0$ and $f > 0$ for the left eye and $F < 0$ and $f < 0$ for the right eye. Finally, considered the minimum of the coefficient of second fundamental form e for extracting of left and right inner eye corner point. The flow diagram of the endocanthion detection process is given in Fig. 8.

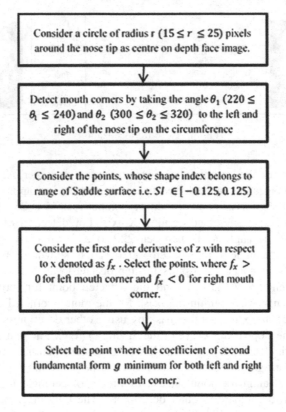

Fig. 7. Flow diagram of Chelion detection

Exocanthions. Exocanthions are the points where the upper and lower outer eyelids meet at outer eye corners. According to [17], first these exocanthion points are detected using the geometrical approach, As per previous, considered a circle of radius r is in between 25 to 35 and two separate angles with respect to X-axis. For detection of the left eye inner corner, considering the angle θ_1 in between 150 to 170 and for right mouth corner, considered the angle θ_2 in between 10 to 30 as shown in Fig. 10(e) and (f). Further, we have used shape descriptors for appropriate localization. Here, the SI belongs to the rut surface $[-0.625, -0.375)$. Also, consider the points, where the first order derivative with respect to x is greater than zero for both left and right outer eye corner. Next, from the resultant points, select those points, where the first-order derivative concerning y is greater than zero for both eyes. Further, clustered the points into two clusters using the k-means clustering technique. Considered points of one cluster, whose center value is minimum. For more optimum localization, considered the Gaussian and Mean curvature, K = 0 and H > 0 for both left and right eye. Finally, considered the minimum of the coefficient of second fundamental form e for extracting of the left and right inner eye corner point. The flow diagram of the exocanthion detection process is given in Fig. 9.

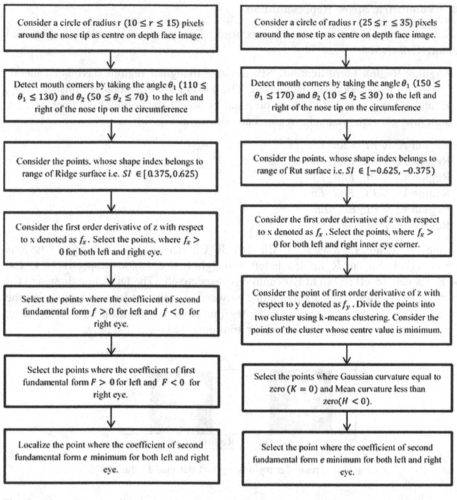

Consider a circle of radius r (10 ≤ r ≤ 15) pixels around the nose tip as centre on depth face image.

Consider a circle of radius r (25 ≤ r ≤ 35) pixels around the nose tip as centre on depth face image.

Detect mouth corners by taking the angle θ_1 (110 ≤ θ_1 ≤ 130) and θ_2 (50 ≤ θ_2 ≤ 70) to the left and right of the nose tip on the circumference

Detect mouth corners by taking the angle θ_1 (150 ≤ θ_1 ≤ 170) and θ_2 (10 ≤ θ_2 ≤ 30) to the left and right of the nose tip on the circumference

Consider the points, whose shape index belongs to range of Ridge surface i.e. $SI \in [0.375, 0.625)$

Consider the points, whose shape index belongs to range of Rut surface i.e. $SI \in [-0.625, -0.375)$

Consider the first order derivative of z with respect to x denoted as f_x. Select the points, where $f_x > 0$ for both left and right eye.

Consider the first order derivative of z with respect to x denoted as f_x. Select the points, where $f_x > 0$ for both left and right inner eye corner.

Select the points where the coefficient of second fundamental form $f > 0$ for left and $f < 0$ for right eye.

Consider the point of first order derivative of z with respect to y denoted as f_y. Divide the points into two cluster using k-means clustering. Consider the points of the cluster whose centre value is minimum.

Select the points where the coefficient of first fundamental form $F > 0$ for left and $F < 0$ for right eye.

Select the points where Gaussian curvature equal to zero ($K = 0$) and Mean curvature less than zero($H < 0$).

Localize the point where the coefficient of second fundamental form e minimum for both left and right eye.

Select the point where the coefficient of second fundamental form e minimum for both left and right eye.

Fig. 8. Flow diagram of Endocanthion detection

Fig. 9. Flow diagram of Exocanthion detection

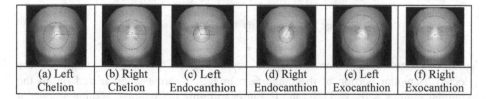

(a) Left Chelion	(b) Right Chelion	(c) Left Endocanthion	(d) Right Endocanthion	(e) Left Exocanthion	(f) Right Exocanthion

Fig. 10. Mouth, inner and outer eye corner localizations

2.4   Volumetric Space Representation

This section is divided into two subsections: Triangular region identification and Volume Calculation.

Triangular Region Identification. Six separate triangular regions have been identified based on three individual landmarks. The total of six triangular regions: A, B, C, D, E, and F have been built, where pronasal is one of the common vertices of each separate triangle. The triangular regions are shown in Fig. 11:

- Region A: ∇PL_ChR_Ch
- Region B: ∇PR_ChR_Ex
- Region C: ∇PR_EnR_Ex
- Region D: ∇PL_EnR_En
- Region E: ∇PL_ExL_En
- Region F: ∇PL_ChL_Ex

Here 'P' denotes Pronasal, L_Ch and R_Ch denote left, and right Chelions respectively, L_En, and R_En denote left, and right Endocanthions respectively, L_Ex, and R_Ex denote left and right Exocanthions respectively. The DDA line drawing [18] algorithm is used for the creation of a line between two distinct landmark points on the range face image. All total, six triangular regions cover the largest portion of the face. Figure 12 shows the coverage area of six triangular regions separately.

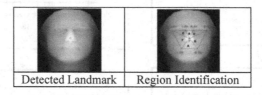

| Detected Landmark | Region Identification |

Fig. 11. Triangular regions against detected landmarks

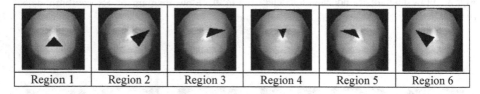

| Region 1 | Region 2 | Region 3 | Region 4 | Region 5 | Region 6 |

Fig. 12. Area of six triangular region

Volume Calculation. Volumetric space representation gives the volume of the selected triangular portion of the range face image. Volume is calculated on all the triangular regions A, B, C, D, E, and F. For calculating the volume, assume a plane at the maximum depth position, i.e., the nose tip of the depth face image. Next, the distance is calculated between the plane and all the pixels of the depth face image, which is shown in Fig. 13(a). The distance between the fitted plane of maximum depth and pixels of the depth image is treated as the density of that pixel. Further, the addition of all pixel densities of any particular region produces the volume of that region. The density calculation of each pixel coordinate generates the inverse range image that is also termed as a density range image, which gives the volume of the image. In this work, we have considered the volume of six selected triangular face region that shown in Fig. 13(b).

Here, the scanline approach of computer graphics is used for separating the volume of each identified triangular region from the density range image. Moreover, choosing the triangular region, three information are to be known to separate the volume of that triangular region:

- Choosing of starting point
- Choose a row-wise or column-wise scanline approach
- Farthest point detection in between the other two points

In our work, we have considered the nose tip or pronasal point of the face as the starting point because this point is common in all the triangular regions. Three cases, according to the x and y positions of the points, occur before applying a scanline approach, as shown in Fig. 14. In the first case, consider the image axis if the x value of the other two points (excluding starting point) are on the same side of Y-axis, i.e., x values of both the points are either increasing or decreasing, and y values of both the points are on different sides of X-axis then considering column-wise scanline approach as shown in Fig. 14(a). Similarly in the second case, when the y value of the other two points (excluding starting point) are on the same side of X-axis, i.e. y values of both the points are either increasing or decreasing, and x values of both the points are on different sides of Y-axis then considering row-wise scanline approach as in Fig. 14(b). Other than these two cases, the third case, when both x and y value is either increasing or decreasing concerning any of the axes, then any approach can be considered as in Fig. 14(c).

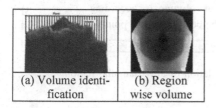

(a) Volume identification	(b) Region wise volume

Fig. 13. Volume calculation on separate regions

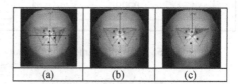

(a)	(b)	(c)

Fig. 14. Different cases of scanline technique

Now, due to aliasing effect, a problem may arise at the time when scanline approach is applied between two different sides of the triangle that is the inequality of the index of the line array, which is not same as x value or y value difference. The flow diagram, in Fig. 15, shows the calculation of the total density value of any triangular region.

The proposed work of volumetric representation can be mathematically justified by volume integral. It can be usually denoted as in Eq. 4.

$$\int_V f(x,y,z)d\tau \tag{4}$$

Here $f(x,y,z)$ denoted as z value of (x, y) location of the negative range image, which is identified as the distance between the highest depth plane and any particular pixel. The function is the summation of all pixels in a particular region. Let a, b and c are the sides of the triangle and A denote area of the circle. Equations 5 and 6 illustrate the volume of the triangular region in this work.

$$A = \sqrt{S(S-a)(S-b)(S-c)} \quad \text{where } S = \frac{(a+b+c)}{2} \tag{5}$$

$$V = \int_{Z_1}^{Z_2} Adz, \quad \text{where } Z_1 \text{ and } Z_2 \text{ are density values} \tag{6}$$

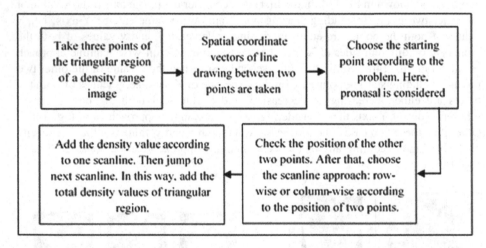

Fig. 15. Flow diagram of total density calculation of any triangular region

2.5   3D Voxel Representation

Voxelization-based 3D object representation is a representation of three-dimensional depth in a different way. Here, we have used volume data to construct 3D voxel along with some geometric characteristic such as area curvature, etc. Voxel [19] denoted by a regular grid in three-dimensional space. The X and Y axes of a 3D voxel correspond to XY-plane of 2D depth images. The volume data denote the Z-axis. In our work, the voxel representation of six defined triangular regions is shown in Fig. 16.

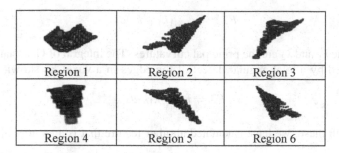

Region 1	Region 2	Region 3
Region 4	Region 5	Region 6

Fig. 16. 3D voxel representation of six triangular regions

2.6 Feature Extraction

Feature extraction is one of the major tasks of the feature-based recognition system. In this work, initially, we have calculated total density or volume of any triangular region, which is a single additive feature for the classification of any triangular region. Next, some of the statistical features [20] are considered from each of the triangular regions. The statistical features have been extracted from the individual volumetric triangular region. There exist various statistical features based on the gray level histogram, gray level co-occurrence matrix (GLCM), edge length distribution, and run length distribution. Here, we have used the first two statistical features, i.e., gray level histogram, gray level co-occurrence matrix-based features. The histogram-based features include mean, variance, standard deviation, skewness, etc. The mean gives the average gray level of each region. The variance calculates the average squared difference from the mean value. The standard deviation provides a variation from the mean value. The skewness gives the asymmetry of the gray levels around the sample mean. Other than histogram-based features, here we have considered one GLCM –based feature, i.e., entropy. The entropy of the image gives the measure of randomness of the gray image levels. In our work, a gray-level value denotes the value of its corresponding volume density.

Next, we have focused on computing the local features of the 3D voxel representation. The local voxel features are calculated from the intersected portion of cubic voxel V and surface S. Here, we have selected the features from a geometrical integral point of view. First, we have considered the three features of intrinsic volume [21] such as the surface area, the integrated mean and Gaussian curvature of the surface. Generally, these features are more significant in pattern recognition problems [22]. The surface area is an additive feature as in Eq. 7, which is calculated by the total area of the surface (corresponding voxel form of size $N \times N \times N$).

$$F1 = \int_{S \cap V} ds \qquad (7)$$

Next, the feature, integral of the Mean curvature is calculated on 3D voxel. This is also defined by a total angular defect of all edges found in the voxel V. It is also an additive feature. Equation 8 illustrates the mathematical expression of this feature.

$$F2 = \int_{U \cap V}(k_1 + k_2)ds \tag{8}$$

Where the k_1 and k_2 are the principal curvatures. The integral of Gaussian curvature is also defined by a total angular defect of the edges in a voxel, as shown in Eq. 9.

$$F3 = \int_{U \cap V} k_1 k_2 ds \tag{9}$$

Other than these, we have calculated another feature that is area normal, as shown in Eq. 10.

$$F4 = \int_{U \cap V} nds \tag{10}$$

Where n is a vector of three dimensions, which is the local normal to the surface. This is also an additive feature. This feature is three dimension vector, whereas the previous three features are one dimension. Compare with previous three features; this feature is orientation dependent. The four distinguish features F1, F2, F3, and F4 are used to create a feature vector for classification. All the feature representation corresponds to a triangular voxel region is shown in Fig. 17 given below.

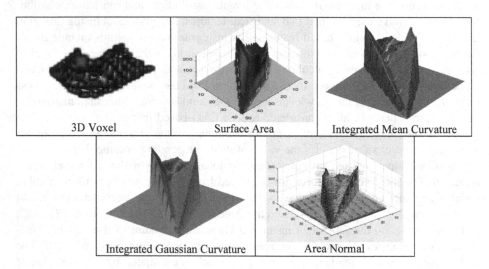

Fig. 17. Surface with different local features

3 Experimental Result and Analysis

The proposed work is applied to range images of three public databases Frav3D, Bosphorus, and GavabDB. The details of these databases are given below in Table 1.

Table 1. Details of input databases

Database	Year of capture	No. of Subjects	No. of scans	Variation
Bosphorus [23]	2008	105(60-M, 45-F)	4652	Pose, expression, occlusion
FRAV3D [24]	2006	106	1696	Pose, expression, illumination
GavabDB [25]	2004	61(45-M, 16-F)	427	Smile, frontal accentuated laugh, frontal random gesture

The different 3D scanners were used to capture the 3D data of these databases. Initially capturing was done as a 3D point cloud and then those were transformed into range images. In the preliminary version of this work [7], we had considered only frontal pose images with expression-variant range images of size 100×100 as input. In this extended version, frontal with different pose-variant and expression-variant range images of the same size as in [7] are considered as input. For pose variant images, at first, the ICP-based technique is used to register the pose variant images. Next, to find the registration accuracy, we have calculated root mean square error (RMSE) between the original frontal face image and the frontal face obtained after registration of an image of the same subject. Table 2 illustrates the registration accuracy based on different thresholds of RMSE results. Next, Fig. 18 shows the identification of proposed landmarks on various registered face images. The accuracy of the landmark localization is shown in Table 3. The table represents the comparative results of localization of seven distinct landmarks in terms of the standard deviation error (SDE) between the present method and method reported by Bagchi et al. with respect to the ground truth. In the experiment, we have considered four neutral, six pose, and two-expression variant face images of FRAV3D, two neutral, six expression, twenty lower face AU (Action Units) (LFAU), five upper face AU (UFAU), two Combined AU (CAU), and fourteen pose variation face images of Bosphorus, and two neutral, two expression, and four pose variation face images of Gavab database. The databases contain different pose variants of images across yaw, roll, pitch, and cross, i.e., the combination of yaw and pitch. Pose variation along X-axis denoted as Roll, along Y-axis denoted as Yaw, along Z-axis denoted as Pitch.

Table 2. Face registration accuracy of three individual datasets

Database	RMSE (Threshold ≤ 5)	RMSE (Threshold ≤ 10)
Frav3D	93	100
Bosphorus	65	90
Gavab	67	90

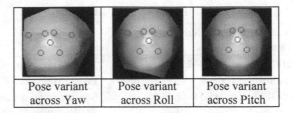

| Pose variant across Yaw | Pose variant across Roll | Pose variant across Pitch |

Fig. 18. Landmarks detection on registered pose variant face

Table 3. Comparison of methods in terms of standard deviation error

Database	Variation	SDE (proposed work)							SDE (Bagchi et al. [16])						
		P	L_Ch	R_Ch	L_En	R_En	L_Ex	R_Ex	P	L_Ch	R_Ch	L_En	R_En	L_Ex	R_Ex
Frav3D	Neutral	0.33	4.24	2.5	3.14	3.78	2.27	3.98	0.35	3.98	2.13	4.36	3.53	3.43	5.66
	Expression	0.5	6.82	4.42	3.14	3.32	1.66	2.54	0.61	6.6	5	3.28	3.67	3.12	3.79
	Pose-Yaw	0.66	8.59	7.64	4.56	4.0	3.12	3.4	0.6	8.11	8.2	5.23	4.5	4.32	4
	Pose-Roll	0.5	5.41	6	3.76	4.23	2.78	2.32	0.52	4.43	4.27	4	3.65	3.22	3.65
	Pose-Pitch	0.78	6.4	5.77	4	4.55	4.1	3.89	0.7	7.12	7	5.6	4.95	4.3	4.75
Bosphorus	Neutral	0.6	3.65	4.72	4.77	4.23	2.3	3.1	0.65	3.2	5.21	5.32	4.78	4.56	4.13
	Expression	0.55	5.41	6	3.28	2.89	4.3	2.98	0.4	5.74	6	4.87	4	5.47	4.23
	CAU	0.32	6.5	5.55	4.1	3.7	2.16	3.67	0.3	6.8	5.28	5.3	4.23	2.35	4.79
	LFAU	0.53	6.3	4.56	3.12	2.59	3.3	2.98	0.6	7.1	4.23	4.22	4	3.9	3.14
	UFAU	0.2	5.33	3.88	2.73	3	5.77	3.21	0.22	5	4.1	3.64	3.11	7.31	4.68
	Pose-Yaw	1.44	7.34	7.66	5.3	4.78	4.22	4.25	0.6	7.5	7.5	5.74	5.1	4.34	4.4
	Pose-Pitch	0.7	6.4	6	5.4	5.02	4.21	4.3	0.76	6.82	6.3	5.53	5.82	4.53	4.6
	Pose-Cross	0.9	6.1	6.59	5.12	5.47	5.62	5.47	0.88	6.32	6.6	5.44	5.32	5.1	5.07
GavabDB	Neutral	0.3	3.86	2.56	2.87	3.7	1.5	2.34	0.32	4	2.44	3.11	3.72	3.2	3.35
	Expression	0.41	4.25	3	5.48	4.71	2.11	3.19	0.38	4.75	4.1	5.7	5.23	4.15	4.63
	Pose-Yaw	1.78	5.13	5.85	5.8	6.12	6.42	6.3	0.67	5.12	6.1	5.87	5.81	6.3	6
	Pose-Pitch	0.59	4.34	4.2	4.23	4.12	5.1	4.76	0.5	5.44	4.3	4.15	4.28	5.05	5.4

The recognition accuracy is calculated on separate triangular regions. The experiment is separated into two parts. First, we have considered only frontal face images as input, including neutral and expression variations as in [7]. Second, we have considered both frontal and registered frontal face images together as input, including neutral and expression variations. The 2-fold cross-validation technique is used, i.e. training, and test sets are equally divided in all cases of the experiment. Now, we have used total volume density as the feature of the individual region.

The recognition accuracy of the system has calculated in two separate ways. Considering volumetric space representation for finding the accuracy in one way, whereas the 3D voxel representation from volumetric space to find the recognition accuracy in different way. Now, the overall time complexity of our method using first approach is $O(N^2)$, where N denotes set of 3D points of 3D face.

Further, we have calculated the recognition accuracy of the whole face by taking average recognition accuracy of all distinct regions. Tables 4 and 5 illustrate the recognition rates of three databases using kNN and SVM classifiers. Next, we have created another feature vector using five distinct statistical features on those triangular regions. The details of the considered statistical features are discussed earlier. The SVM and kNN classifiers were used for classification in the specific region. Tables 6 and 7 illustrate the recognition accuracy using the statistical features.

Table 4. The recognition accuracy of three input databases using kNN classifier based on total volume density.

Database	Input	Region A	Region B	Region C	Region D	Region E	Region F	Average
Frav3D	Frontal	94.4	93.8	94	94.9	94.6	94	94.28
	All (Frontal+Registered frontal)	92.5	91.87	92.48	91.89	92	92.4	92.19
Bosphorus	Frontal	95.88	95	94.77	95.45	95	95.2	95.21
	All (Frontal+Registered frontal)	92.1	93.4	92.47	92.3	93.42	92	92.61
GavabDB	Frontal	91.68	90.3	91	90.6	91.4	90	90.83
	All (Frontal+Registered frontal)	85.68	85.9	86.67	85	84.84	84	85.34

Table 5. The recognition accuracy of three input databases using SVM classifier based on total volume density.

Database	Input	Region A	Region B	Region C	Region D	Region E	Region F	Average
Frav3D	Frontal	95.58	95.3	96	95.9	95.78	95	95.59
	All (Frontal+Registered frontal)	92.33	93.12	93.5	93.14	93.36	92.49	92.9
Bosphorus	Frontal	96.87	96	96.7	96	96.5	96.2	96.37
	All (Frontal+Registered frontal)	92.8	92.56	91.9	92.48	92.37	92.7	92.46
GavabDB	Frontal	93.68	91.3	92	93.8	91.89	92.4	92.51
	All (Frontal+Registered frontal)	87.36	86.4	86.91	87	86.1	86.54	86.71

Table 6. The recognition accuracy of three input databases using kNN classifier based on statistical features.

Database	Input	Region A	Region B	Region C	Region D	Region E	Region F	Average
Frav3D	Frontal	94	93.38	94.19	94.87	94.72	93.89	94.17
	All (Frontal+Registered frontal)	92.24	92.1	92.4	92.8	92.47	92.38	92.39
Bosphorus	Frontal	96	94.85	94.97	95	95.43	95.58	95.30
	All (Frontal+Registered frontal)	92.41	92.5	91.89	92.12	92.7	92.78	92.4
GavabDB	Frontal	91.5	90.45	91.4	90.87	90.5	90.2	90.82
	All (Frontal+Registered frontal)	85.3	85.7	86.75	85.34	85.5	84.15	85.4

Table 7. The recognition accuracy of three input databases using SVM classifier based on statistical features.

Database	Input	Region A	Region B	Region C	Region D	Region E	Region F	Average
Frav3D	Frontal	95.5	95	96.12	95.78	95.8	95.2	95.56
	All (Frontal+Registered frontal)	92.34	92.1	93.22	93	93.17	92	92.63
Bosphorus	Frontal	96.6	96.2	96.61	96	96.22	96.34	96.32
	All (Frontal+Registered frontal)	92.59	92.2	92.7	91.8	92.28	92.4	92.32
GavabDB	Frontal	93.27	91	92	94.1	91.91	92.36	92.44
	All (Frontal+Registered frontal)	87	85.4	86	87.64	85.8	86.62	86.41

In the voxel-based representation, the classification process is applied to the extracted geometric feature vector $[F1\,F2\,F3\,F4]$. As per previous, 2-fold cross validation with two types of inputs: frontal, and frontal with registered are considered for the experiment. The SVM classifier is used for classification on the three input 3D face datasets. Other than previous approach for recognition, the time complexity of this approach is $O(N^3)$, here N denotes the size of 3D voxel. Table 8 illustrates the face recognition accuracy using voxel-based features.

Table 8. The recognition accuracy of three input databases based on geometrical voxel features

Database	Input	Region A	Region B	Region C	Region D	Region E	Region F	Average
Frav3D	Frontal	92	91.69	92.6	91.35	92.79	92.11	92.09
	All (Frontal+Registered frontal)	88.32	89.65	86.8	88.45	87.7	88.37	88.2
Bosphorus	Frontal	93.98	93.14	94	94.31	93.22	93.4	93.67
	All (Frontal+Registered frontal)	89.2	90.12	89.33	91	90.3	90.65	90.1
GavabDB	Frontal	90.23	89.67	90.1	89.8	89.3	89.12	89.7
	All (Frontal+Registered frontal)	85.9	85.73	85.44	86.3	84.46	85	85.47

4   Comparative Study and Analysis

In our proposed work, we have used three databases: Frav3D, Bosphorus, and Gavab as input databases. In this section, our proposed method compare with previous other tasks of the three different databases. Table 9, 10, and 11 illustrate the comparative study.

Table 9. Comparison of recognition performance of the proposed method with some other methods on the Frav3D database

Methods	Accuracy (%)	Reference
Curvature analysis + SVD + ANN (Classification on the whole face) (Input: Only frontal with expression and Illumination variation)	86.51	Ganguly et al. [26]
Curvature analysis + SVD + ANN (Classification on the whole face) (Input: Frontal with expression and illumination variation + Non-frontal)	76.08	Ganguly et al. [26]
LBP + HOG + kNN (Region based classification) (Input: Only frontal with expression and illumination variation)	88.86	Dutta et al. [11]
Geodesic texture wrapping + Euclidean-based classification (Classification on the whole face) (Only pose variation)	90.3	Hajati et al. [27]
Proposed method (Triangular representation + Volume calculation + 3D voxel representation + Geometric features + SVM) (Input: Only frontal with expression and illumination variation)	**92.09**	This paper
Proposed method (Triangular representation + Volume calculation + 3D voxel representation + Geometric features + SVM) (Input: Frontal with expression and illumination variation + Non-frontal)	**88.2**	This paper
ICP based registration + Surface Normal + KPCA (Input: Only pose variation)	92.25	Bagchi et al [28]
Proposed method (Triangular representation + Volume calculation + Statistical feature + kNN) (Input: Only frontal with expression and illumination variation)	**94.17**	This Paper
Proposed method (Triangular representation + Volume calculation + Statistical feature + kNN) (Input: Frontal with expression and illumination variation + Non-frontal)	**92.39**	This paper
Triangular representaion + Volume calculation + Total density feature + kNN (Input: Only frontal with expression and illumination variation)	94.28	Dutta et al. [7]
Proposed method (Triangular representation+ Volume calculation + Total density feature + kNN) (Input: Only frontal with expression and illumination variation)	**94.19**	This paper
DWT + DCT + PCA + Euclidean distance classifier (Input: Only pose variation)	94.50	Naveen et al. [29]
Proposed method (Triangular representation + Volume calculation + Statistical feature + SVM) (Input: Only frontal with expression and illumination variation)	**95.56**	This paper
Proposed method (Triangular representation + Volume calculation + Statistical feature + SVM) (Input: Frontal with expression and illumination variation + Non-frontal)	**92.63**	This paper
Triangular representation + Volume calculation + Total density feature + SVM (Input: Only frontal with expression and illumination variation)	95.59	Dutta et al. [7]
Proposed method (Triangular representation + Volume calculation + Total density feature + SVM) (Input: Frontal with expression and illumination variation + Non-frontal)	**92.9**	This paper

Table 10. Comparison of recognition performance of the proposed method with some other methods on the Bosphorus database

Methods	Accuracy(%)	Reference
ICP based recognition (Input: Frontal with expression variation)	94.10	Dibeklioglu et al. [30]
ICP based recognition (Input: Only Pose variation)	79.41	Dibeklioglu et al. [30]
Proposed Method (Triangular representation + Volume calculation + 3D voxel representation + Geometric features + SVM) (Input: Frontal with expression variation)	**93.67**	This paper
Proposed Method (Triangular representation + Volume calculation + 3D voxel representation + Geometric features + SVM) (Input: Frontal with expression variation + Non-frontal)	**90.1**	This paper
Triangular representation + Volume calculation + Total density feature + kNN (Input: Frontal with expression variation)	95.21	Dutta et al. [7]
Proposed Method (Triangular representation + Volume calculation + Total density feature + kNN) (Input: Frontal with expression variation + Non-frontal)	**92.61**	This paper
Proposed method (Triangular representation + Volume calculation + Statistical feature + kNN) (Input: Frontal with expression variation)	**95.3**	This paper
Proposed method (Triangular representation + Volume calculation + Statistical feature + kNN) (Input: Frontal with expression variation + Non-frontal)	**92.4**	This paper
ICP based holistic approach + Maximum likelihood classifier (Frontal with expression variation)	95.87	Alyuz et al. [31]
ICP based holistic approach + Maximum likelihood classifier (Frontal with occlusion variation)	94.12	Alyuz et al. [31]
ICP based registration + Surface Normal + KPCA (Input: Only pose variation)	96.25	Bagchi et al. [28]
Proposed method (Triangular representation + Volume calculation + Statistical feature + SVM) (Input: Frontal with expression variation)	**96.32**	This paper
Proposed method (Triangular representation + Volume calculation + Statistical feature + SVM) (Input: Frontal with expression variation + Non-frontal)	**92.32**	This paper
Triangular representation + Volume calculation+ Total density feature + SVM (Input: Frontal with expression variation)	96.37	Dutta et al. [7]
Proposed method (Triangular representation + Volume calculation + Total density feature + SVM) (Input: Frontal with expression variation + Non-frontal)	**92.46**	This paper

Table 11. Comparison of recognition performance of the proposed method with some other methods on the Gavab database

Methods	Accuracy(%)	Reference
Mean and Gaussian curvature based Segmentation (Input: Frontal with expression variation + Non-frontal)	77.9	Moreno et al. [32]
Proposed method (Triangular representation + Volume calculation + 3D voxel representation + Geometric features + SVM) (Input: Frontal with expression variation)	**89.7**	This Paper
Proposed method (Triangular representation + Volume calculation + 3D voxel representation + Geometric features + SVM) (Input: Frontal with expression variation + Non-frontal)	**85.47**	This Paper
Geometrical feature + PCA vs. SVM classification (Input: Frontal with expression variation)	90.16	Moreno et al. [3]
Geometrical feature + PCA vs. SVM classification (Input: Pose variation)	77.9	Moreno et al. [3]
Proposed method (Triangular representation + Volume calculation + Statistical feature + κNN) (Input: Frontal with expression variation)	**90.82**	This Paper
Proposed method (Triangular representation + Volume calculation + Statistical feature + κNN) (Input: Frontal with expression variation + Non-frontal)	**85.4**	This Paper
Triangular representation + Volume calculation + Total density feature + κNN (Input: Frontal with expression variation)	90.83	Dutta et al. [7]
Proposed method (Triangular representation + Volume calculation + Total density feature + κNN) (Input: Frontal with expression variation + Non-frontal)	**85.34**	This paper
2DPCA + SVM classifier (Input: Frontal with expression variation)	91	Mousavi et al. [33]
Proposed method (Triangular representation + Volume calculation + Statistical feature + SVM) (Input: Frontal with expression variation)	**92.44**	This paper
Proposed method (Triangular representation + Volume calculation + Statistical feature + SVM) (Input: Frontal with expression variation + Non-frontal)	**86.41**	This paper
Triangular representation + Volume calculation + Total density feature + SVM (Input: Frontal with expression variation)	92.51	Dutta et al. [7]
Proposed method (Triangular representation + Volume calculation + Total density feature + SVM) (Input: Frontal with expression variation + Non-frontal)	**86.71**	This paper

In 2014, the authors developed a 3D face recognition system [26] using a fusion of different 3D curvatures. They have used all variation including pose, expression, and illumination face images of Frav3D database. Compared to our work, they have used an elliptical cropped face from the original image of size 100×100, which he face area is too large compared to our consecutive triangular face portions. Next, they have to use singular value decomposition for feature extraction. For classification, minimum of ten features for creating a feature vector, whereas, in our work, we have considered six elements to create a feature vector. Further, the result of their work improve by considering more than ten features, but in our work, the numbers of features are the same to create a feature vector.

In the year 2015, a 3D face recognition system [28] has been proposed by extracting feature from 2.5D depth face images based on surface normal and Kernel principal component analysis (KPCA). They are also used small pose variations of 3D images of Frav3D, Bosphorus, and Gavab databases as inputs. Compare to our proposed work, and they have considered elliptical cropped face image that consists all the significant portions of a human face, whereas we have considered 1/6th portion with respect to whole face region for generating consecutive six triangular areas. Next, we have considered a feature vector of six individual features for classification, whereas they are used a minimum of 10 features for creating a feature vector. Also, most of the experiments of their work give a better result, when the number of is more than 10. So, their system is unable to produce better recognition accuracy using a small feature set.

In 2016, the authors developed a 3D face recognition system [17] differently. First, seven landmark points have been detected innovatively. Next, the face recognition system has been developed by creating a feature vector by calculating Euclidean distances between each other landmark. A feature vector of 21 numbers of features is used for classification; compare to our work, the feature set is large.

Next, in 2018, a 3D face recognition system [34] has been developed based on a modified local binary pattern (LBP) technique named as TR-LBP. The authors are developed a modified LBP from the shape index values of 2.5D depth images. The original input depth face images of Frav3D and Bosphorus are used for feature extraction followed by classification. The entropy-based 16 numbers of features are used to create a feature vector. Compare to our work, and they are used more numbers of features to create feature vectors. Another limitation of their work is that only frontal with expression and illumination variant images are used as inputs, whereas in our work, we have considered pose variation with expression and illumination variant face images.

From the discussion, considering less number of features for classification in our recognition system takes less amount of time for matching and classification. Also, in some of the work, considering more number of features may give better recognition result. Whereas, our feature set is fixed and definite. Also, the considering face area for feature extraction is too less compared to other proposed work, so there is no need for full-face for recognition.

As per our preliminary work [7], the system acquired 94.28%, 95.21%, and 90.83% using kNN classifier and 95.59%, 96.37%, and 92.51% using SVM classifier by considering only frontal pose with neutral and expression variation face images of Frav3D, Bosphorus, and Gavab databases. The total volume or density calculation was considered in that work [7]. Besides using the feature of previous work, here we have also considered statistical features. Using the statistical features, the system acquires 94.17%, 95.3%, and 90.82% using kNN classifier and 95.56%, 96.32%, and 92.44% using SVM classifier. In this extended work, frontal as well as pose with expression variant depth face images are considered as inputs, and the accuracies are shown in

Tables 4, 5, 6 and 7. On the other side, the newly proposed voxel representation-based classification system acquired 92.09%, 93.67%, and 89.7% respectively using SVM classifier. The present method does not consider the entire face image as it is done by most of the methods, and hence, the size of the feature vectors is less compared to other works. Therefore, the present method has the advantage of the small length feature vector and higher recognition accuracy. Compare to our preliminary work [7], here we have used other different features in both 2.5D and 3D representation with features of previous work, to prove the correctness of the system. Also in [7], we had considered only frontal face depth face as input, whereas, in this proposed method, it works with a variant of images such as neutral, pose, and expression variants, which prove that the system is generalized for different types of input.

We have considered three different feature extraction techniques, such as total density features, statistical features, and 3D voxel-based features. Within these three, the total density feature set and the statistical feature set are classified using two classifiers: KNN and SVM. The only SVM classifier is used for classification of 3D voxel-based features. According to the comparison Table 9 of Frav3D database, considering only frontal images, our method gives better result compare to the work of [26] and [11] on all three cases of features with different classifiers. Next, considering all variation of faces (frontal and registered non-frontal images with expression and illumination variation), three examples of features of our method gives better accuracy compared to the works of [26] and [27]. The result of [28] with respect to pose variant images is lower when statistical features of our method are classified using SVM and KNN classifier.

According to Table 10, considering the Bosphorus database with frontal pose including expression and illumination variation faces, the 3D voxel-based features gives low recognition rate compare with the work [30], whereas in the same scenario, the total density and statistical features give better result using both SVM and KNN classifier. Classifying the statistical features of our work with SVM classifier in case of the frontal pose with expression variation faces, our work gives better accuracy compared with the work of [31], whereas in the same scenario with kNN classifier give low accuracy. Considering total density as feature set with SVM classifier, our work provides better result compare with the work of [31].

From Table 11, in case of the frontal and non-frontal pose with expression variation of Gavab database, all three cases of feature set give better result compare with the work of [32]. In the scenario, when considering frontal face with expression variation images, the work of [3] gives low accuracy compare with our three cases of feature set for two different classifiers. The similar scenario of the previous discussion, the work of [33] provides better result compare with our work with 3D voxel-based feature with SVM classifier, also total density and statistical feature using kNN classifier whereas it allows for low accuracy when considering total density feature or statistical feature using SVM classifier.

5 Conclusion

In this work, the 3D depth face has been represented in volumetric space. Initially, some significant landmarks are identified. Then the triangular region has been identified by calculating line using a DDA line drawing algorithm between three distinct landmarks, where nose tip is a common landmark point in all triangular regions. Next, the volume has been calculated by assuming a plane at the highest depth position depth face image. According to the publishing work, the total density of the volume space of all distinct regions is used to construct feature vectors for classification.

Further, some of the statistical features are also calculated on the individual triangular region for creating a new feature vector. Overall it can be concluded that the volume representation of range images is a new approach in the 3D domain. The system is computationally efficient for recognition with high recognition accuracy.

The work introduces a new representation: 3D voxel that has been constructed from the volume data of volumetric depth image. The 3D voxel representation is also used for 3D face classification in a different way. We have considered some geometrical local features from voxel representation. As defined, the geometrical voxel features play a major role in face classification. Working with 2D depth image or 2.5D image as well as 3D voxel gives us a robust system from input perspective.

Frontal and non-frontal faces, including neutral and expression variant range images of three well-known databases, are used as input to this system. The non-frontal images are the pose oriented images according to X, Y, and Z-axes. The pose variant images are used after registration using ICP technique.

Acknowledgement. The first author is grateful to Ministry of Electronics and Information Technology (MeitY), Govt. of India for the grant of Visvesvaraya doctorate fellowship award. The authors are also thankful to CMATER laboratory of the Department of Computer Science and Engineering, Jadavpur University, Kolkata, India for providing the necessary infrastructure for this work.

References

1. Abate, F., Nappi, M., Riccio, D., Sabatino, G.: 2D and 3D face recognition: a survey. Image Inf. Control **28**(14), 1885–1906 (2007)
2. Gervei, O., Ayatollahi, A., Gervei, N.: 3D face recognition using modified PCA methods. World Acad. Sci. Eng. Technol. **4**(39), 264 (2010)
3. Moreno, A.B., Sanchez, A., Velez, J.F., Diaz, J.: Face recognition using 3D local geometrical features: PCA vs. SVM. In: Proceedings of the ISPA, pp. 185–190 (2005)
4. Heseltine, T., Pears, N., Austin, J.: Three-dimensional face recognition: a fisher surface approach. In: Proceedings of the ICIAR, pp. 684–691 (2008)
5. Hesher, C., Srivastava, A., Erlebacher, G.: A novel technique for face recognition using range imaging. In: Proceedings of the Seventh International Symposium on Signal Processing and Its Applications, pp. 201–204 (2003)
6. Soltanpour, S., Boufama, B., Wu, Q.M.J.: A survey of local feature methods for 3D face recognition. Pattern Recognit. **72**, 391–406 (2017)

7. Dutta, K., Bhattacharjee, D., Nasipuri, M.: 3D face recognition based on volumetric representation of range image. In Chaki, R., Cortesi, A., Saeed, K., Chaki, N. (eds.) Advance Computing and Systems for Security, Advance in Intelligent Systems and Computing, vol. 883, pp. 175–189 (2019)
8. Ganguly, S., Bhattacharjee, D., Nasipuri, M.: 3D face recognition from complement component range face images. In: IEEE International Conference on Computer Graphics, Vision and Information Security (CGVIS), pp. 275–278 (2015)
9. Moreno, A.B., Sanchez, A., Velez, J.F.: Voxel-based 3D face representations for recognition. In: 12th International Workshop on Systems, Signals and Image Processing, pp. 285–289 (2005)
10. Shekar, B.H., Harivinod, N., Kumara, M.S., Holla, K.R.: 3D face recognition using significant point based SULD descriptor. In: International Conference on Recent Trends in Information Technology (ICRTIT), Chennai, Tamil Nadu, India, pp. 981–986 (2011)
11. Dutta, K., Bhattacharjee, D., Nasipuri, M.: Expression and occlusion invariant 3D face recognition based on region classifier. In: 1st International Conference on Information Technology, Information Systems and Electrical Engineering (ICITISEE), pp. 99–104. (2016)
12. Ganguly, S., Bhattacharjee, D., Nasipuri, M.: 2.5D face images: acquisition, processing and application. In: ICC 2014 - Computer Networks and Security, pp. 36–44 (2014)
13. Yin, L., Wang, R., Neuvo, Y.: Weighted median filters: a tutorial. IEEE Trans. Circuits Syst.-11: Analog. Digit. Signal Process. 43(3), 157–192 (1996)
14. Besl, P.J., McKay, N.D.: A method for registration of 3-D shapes. IEEE Trans. Pattern Anal. Mach. Intell. (T-PAMI) 14(2), pp. 239–256 (1992)
15. Ahdid, R., Taifi, K., Safi, S., Manaut, B.: A survey on facial features points detection techniques and approaches. Int. J. Comput. Electr. Autom. Control Inf. Eng. 10(8), 1566–1573 (2016)
16. Vezzetti, E., Marcolin, F.: Geometry-based 3D face morphology analysis: soft-tissue landmark formalization. Multimed. Tools Appl. 68(3), 895–929 (2014)
17. Bagchi, P., Bhattacharjee, D., Nasipuri, M.: A robust analysis, detection and recognition of facial features in 2.5D images. Multimed. Tools Appl. 75(18), 11059–11096 (2016)
18. DDA line drawing. http://www.geeksforgeeks.org/dda-line-generation-algorithm-computer-graphics/
19. Shin, D., Fowlkes, C.C., Hoiem, D.: Pixels, voxels, and views: a study of shape representations for single view 3D object shape prediction. CVPR (2018)
20. Vijayarekha, K.: Feature Extraction. https://nptel.ac.in/courses/117106100/Module%209/Lecture%204/LECTURE%204.pdf
21. Santalo, L.A..: Integral Geometry and Geometric Probability. Cambridge Mathematical Library. Beijing World Publishing Corporation (BJWPC) (2004)
22. Mecke, K.R.: Additivity, convexity, and beyond: applications of Minkowski functionals in statistical physics. In: Mecke, K.R., Stoyan, D. (eds.) Statistical Physics and Spatial Statistics. LNP, vol. 554, pp. 111–184. Springer, Berlin (2000). https://doi.org/10.1007/3-540-45043-2_6
23. Bosphorus. http://bosphorus.ee.boun.edu.tr/default.aspx
24. FRAV3D. http://www.frav.es/databases
25. GavabDB. http://gavab.escet.urjc.es/recursos_en.html
26. Ganguly, S., Bhattacharjee, D., Nasipuri, M.: 3D face recognition from range images based on curvature analysis. ICTACT J. Image Video Process. 4(3), 748–753 (2014)
27. Hajati, F. Gao, Y.: Pose-invariant 2.5D face recognition using geodesic texture warping. In: 11th International Conference on Control, Automation, Robotics and Vision Singapore, pp. 1837–1841 (2010)

28. Bagchi, P., Bhattacharjee, D., Nasipuri, M.: 3D face recognition using surface normals. In: TENCON 2015 - 2015 IEEE Region 10 Conference, pp. 1–6 (2015)
29. Naveen, S., Moni, R.S.: A robust novel method for face recognition from 2D depth images using DWT and DCT fusion. In: International Conference on Information and Communication Technologies (ICICT), Elsevier, pp. 1518–1528 (2014)
30. Dibeklioğlu, H., Gökberk, B., Akarun, L.: Nasal region-based 3D face recognition under pose and expression variations. In: Tistarelli, M., Nixon, M.S. (eds.) ICB 2009. LNCS, vol. 5558, pp. 309–318. Springer, Heidelberg (2009). https://doi.org/10.1007/978-3-642-01793-3_32
31. Alyuz, N., Gokberk, B., Akarun, L.: A 3D face recognition system for expression and occlusion invariance. In: BTAS 2008: Proceedings of the IEEE Second International Conference on Biometrics Theory, Applications and Systems, Arlington, Virginia, USA (2008)
32. Moreno, A.B., Sanchez, A., Velez, J.F., Diaz, J.: Face recognition using 3D surface-extracted descriptors. In: Irish Machine Vision and Image Processing Conference (2003)
33. Mousavi, M.H., Faez, K., Asghari, A.: Three dimensional face recognition using SVM classifier. In: Seventh IEEE/ACIS International Conference on Computer and Information Science, Portland, pp. 208–213 (2008)
34. Dutta, K., Bhattacharjee, D., Nasipuri, M.: TR-LBP: a modified local binary pattern-based technique for 3D face recognition. In: Fifth International Conference on Emerging Applications of Information Technology (EAIT) (2018)

Combining Merkle Hash Tree and Chaotic Cryptography for Secure Data Fusion in IoT

Nashreen Nesa[✉] and Indrajit Banerjee

Department of Information Technology, Indian Institute of Engineering Science and Technology, Shibpur, Howrah 711103, West Bengal, India
{nashreennesa.rs2016,ibanerjee}@it.iiests.ac.in
http://www.iiests.ac.in

Abstract. With the wide applicability of sensors in our daily lives, security has become one of the primary concerns in an Internet of Things (IoT) environment. Particularly, user's privacy and unauthorized access to sensitive information needs to be kept in mind while designing security algorithms. This paper puts forward a security protocol that integrates authentication of the deployed IoT devices and encryption of the generated data. We have modified the well-known Merkle Hash Tree to adapt to an IoT environment for authenticating the devices and utilized the concepts of Chaos theory for developing the encryption algorithm. The use of chaos in cryptography are known to satisfy the basic requirements of the cryptosystem such as, high sensitivity, high computational speed and high security. In addition, we have proposed a chaotic map named Quadratic Sinusoidal Map which exhibits better array of chaotic regime when compared to the traditional quadratic map. The security analysis demonstrate that the proposed protocol is simple having low computational requirements, has strong security capabilities and highly resilient to security attacks.

Keywords: Chaos theory · Merkle Hash Tree · IoT · Security · Encryption

1 Introduction

Past researches in the field of Internet security are limited to traditional internet, but with the recent advancement of IoT technologies, these solutions need to be refined so as to cater to the specific needs of IoT [1,2]. Security algorithms for IoT applications should be such that it ensures source authentication, confidentiality, data integrity and resistance against attacks [3]. There is no denying the fact that every smart object in an IoT environment carries the potential of becoming the entry point of malicious activity. Essentially in IoT, security is of paramount importance since this emerging technology revolution entirely depends on the acceptability of its customers. As a result, fusion of data, that

© Springer-Verlag GmbH Germany, part of Springer Nature 2020
M. L. Gavrilova et al. (Eds.): Trans. on Comput. Sci. XXXV, LNCS 11960, pp. 85–105, 2020.
https://doi.org/10.1007/978-3-662-61092-3_5

is the basis upon which the critical decisions are taken, becomes more challenging if end-to-end security between a sender and a receiver is not ensured. In IoT applications, nodes are constantly communicating highly sensitive data and due to this, such data are vulnerable to security attacks. Authentication is a prerequisite and the most important requirement for secure communication in IoT applications since the communicating devices are prone to security attacks. It is essential for every communicating device in the network to verify its identity so that no unauthorized device can take part in communication [4]. Data confidentiality and integrity are also equally important since malicious alteration of sensor data may result in life-threatening consequences especially in critical IoT applications such as healthcare [5–7]. Data confidentiality that is achieved through encryption of the sensed data is vital in order to ensure that no unwarranted disclosure of sensitive information is possible [8]. Thus, in our work, we have proposed a data fusion approach that ensures the security of the devices as well as the data generated to form useful, reliable and secured result. In our proposed security scheme, Chaos theory is used for encryption and decryption of the data and Merkle Hash Tree for device authentication. Since IoT devices have in-built radio frequency identification (RFID) tags for device identification, we have found its application in our work where it is used for the purpose of authentication by exploiting its uniqueness to serve our purpose. Moreover, we have proposed a sinusoidal chaotic map that is used for encryption; the initial condition and the control parameters of which are produced from the Merkle Hash Tree that forms the basis upon which the maps are created. Our work uses lightweight computation modules, such as one-way hash functions and bitwise exclusive-or operation, for designing the secure data fusion protocol. Besides being a lightweight computation tool, the use of hash operations also preserves anonymity since the hash values are impossible to regenerate. In secure communication, the receiver should also have the provision to examine whether the message has been altered during transmission. For this purpose, this paper adopts a simple mechanism for integrity checking by padding the number of zeros in the original ciphertext. Although researchers have investigated the concept of designing cryptosystem based on chaotic maps in the past, but to the best of our knowledge, this is the first attempt at combining Merkle hash Tree with a novel chaotic map in order to achieve security. Specifically, our contributions are listed as follows:

- First, we present an authentication scheme based on the Merkle hash tree technique where the hash values of the leaves are calculated on the unique RFID tags attached to IoT devices.
- Second, we propose a novel Quadratic Sinusoidal chaotic map whose dynamical characteristic properties are studied and confirmed to belong to the chaotic community.
- Third, an efficient data fusion protocol that is based on the Merkle Hash Tree and the chaos theory is proposed. The Merkle Hash Tree generates the initial conditions and the control parameters of the chaotic map that are used for

Table 1. Common Notations used in this work

Notation	Description
D_i	i^{th} IoT device
n	Number of devices in the network
TC	Trusted Data Fusion Centre
l	Number of levels in MHT
$\phi_{i,j}$	Merkle hash assignment of the ith node at the jth Level of MHT
H(.)	Secure one-way Hash operation i.e. SHA-1
$\oplus$	XOR operation
$\|\|$	Concatenation operation
l	Number of levels in Merkle Hash Tree
θ	Merkle Hash Path
$\mathcal{S}$	Pre-shared Session key
$\mathcal{K}$	Initial condition of the map; also serves as the key
Itr	Number of iterations in the map
$\mathcal{P}_i^t$	Plaintext from the ith device at the tth instant
$\mathcal{C}_i^t$	Ciphertext from the ith device at the tth instant
$N^{(0)}$	Zero count in ciphertext $\mathcal{C}_i^t$
$Cipher$	Final ciphertext after appending $N^{(0)}$ i.e., $\mathcal{C}_i^t\|N^{(0)}$

encryption/decryption of messages. After which they are effectively fused to derive the intended result.

– Lastly, extensive security analysis indicates that the proposed scheme can resist all kinds of attacks in addition to ensuring data integrity, confidentiality and authenticity.

The remainder of the paper is organized as follows: a related study on the recent trends in research is presented in Sect. 2, followed by the introduction of concepts of Merkle Hash Tree with its key definitions in Sect. 3. Details about our proposed Modified Sinusoidal Quadratic map is presented in Sect. 4. Next, in Sect. 5, a description of all the phases in our proposed architecture is given. Section 6 describes an experimental scenario of our proposed algorithm for easy understanding, followed by the security analysis of the algorithms in Sect. 7. Finally, the paper is concluded with its ending remarks in Sect. 8.

2 Related Works

Developing security solutions that fulfils the specific requirements of IoT environment is a challenging task and currently a lot of researchers are focussing on this domain. Owing to the distributed network of the IoT devices, security solutions are often integrated with cloud servers and could computing technologies

[9–11]. Specifically, in [9], a mutual authentication scheme is presented based on Elliptic Curve Cryptography (ECC) for secure communication between devices and cloud servers. The proposed protocol has been verified using AVISPA tool to be highly efficient with low computational cost. Since, conventional cryptography solutions are not applicable for IoT applications and schemes developed for IoT must be lightweight. Owing to this requirement, the authors in [10] propose a light weight authentication protocol for IoT enabled devices. For mutual authentication, BAN logic has been used and the protocol was simulated using AVISPA software. Another closely related work is presented in [11], where a robust authentication scheme for resource-constraint IoT devices with cloud assistance is designed. The proposed scheme is lightweight since only cryptographic modules such as one-way hash functions and XOR operations are used. Moreover, security in terms anomaly detection for specific applications can be found in the available literature. For instance, in [12], the authors have introduced a secured IoT-based traffic system with intelligence that is capable of analyzing traffic data into good or bad using Support Vector Machine (SVM). The system was implemented using Raspberry Pi3 and Scikit. Similar to this work, the authors in [13] used four machine learning algorithms for detecting forest fire based on real-world dataset. In addition, an IoT architecture is designed that distinguishes cases when the sensors are faulty and when a fire is detected. Subsequent measures and alert through the use of IoT technologies have also been incorporated. Another closely related work is that presented in [14] where the authors proposed a sequence based learning algorithm to detect inconsistent data in IoT devices. The proposed algorithm was tested for both faulty node detection referred to as "Error" and any abnormal activity known as "Event" using three different real-life datasets. Inspired by the human immunity system under pathogenic attacks, a bio-inspired security solution is proposed in [15]. The proposed solution leverage supervised k-mean-based learning to first distinguish the faulty nodes from the good ones, after which it introduces virtual antibodies to deactivate the fraudulent nodes in the system. Owing to the limited computational capabilities of IoT objects and with the intention of helping designers estimate the cost of implementing security solutions, [16] is proposed. Here, the authors presented a formal framework based on the process calculus IoT-LySa [17] to ascertain a trade-off between security solutions and their cost.

Our work is an extension of our previous work [18] that uses chaotic cryptography for developing an efficient light-weight encryption algorithm. Chaos is a popular theory in numerous natural and laboratory systems encompassing several scientific and research areas as a result of which there is a rich body of literature dedicated to chaos theory. A review on recent enhancements of traditional data encryption procedures is given in [19].

Particularly, the work in [20] relates to body area networks (BANs) application where the authors used image encryption for testing under the assumption that a significant part of sensor data are of images. An efficient flood forecasting model is presented in by studying the data from an area in Brazil through a WSN network. The data was modelled using machine learning techniques and chaos theory. Moreover, a lot of applications have adopted chaos theory for either encryption, detection or authentication [21,22] ranging from pipe leakage detection [23], flood forecasting [24,25], Iris recognition [26], network traffic forecasting [27] to name a few.

In this work we have used the popular Merkle Hash tree (MHT) as an authentication algorithm. MHT has been extensively adopted by researchers in both IoT [28–30] and non-IoT [31–34] related fields. Focussing on MHT, the authors in [28] proposed an authentication scheme for securing smart grid communication. The authentication protocol is based on Merkle Hash Tree where the authors demonstrated that the proposed scheme incurs less computation cost compared with RSA-authentication mechanisms. Security analysis presented by the authors shows that it can resist replay attack, message injection attack, message analysis attack, and the message modification attack. However, not much is discussed about providing integrity and confidentiality in the process of communication. Next, in [29] also proposed a Neighborhood Area Network (NAN) for authenticating power usage power data in smart grids. The authors incorporated digital signature schemes for fault tolerance. In addition, fault diagnosis schemes are also deployed to pinpoint the errors and reduce the computational and communication load in the system. Another work involving smart grids is presented in [30] to provide mutual authentication authenticate between smart meters and the utility servers. A key management protocol is also presented and the whole system is capable to resisting numerous cryptographic attacks.

As can be observed from the available literature both MHT and Chaos theory are highly efficient standalone secularity tools, the combination of which has never been attempted before. Therefore, the need for developing a security algorithm that amalgamates the advantages of both the theories, motivated this research work. Even though a lot of research have been done on security in IoT, to the best of our knowledge, this paper is the first attempt at combining Merkle Hash Tree and chaotic cryptography for developing a security protocol for IoT environment. Preliminary definitions and notions of both MHT and Chaos theory are briefly discussed next.

3 Merkle Hash Tree

Merkle Hash Tree (MHT) was first introduced in 1989 by Merkle [35] and has since then been used for verification and integrity checking by various applications. It is a popular technique among the Git and Bitcoin community for authenticating users. MHTs have mostly been used as authentication schemes [28,31]. A typical MHT is a binary tree in which the nodes of the tree are simple hash values. The nodes at the lowest level (leaf nodes) could be an arbitrary hash values or the hash values generated from pseudorandom numbers whereas the nodes at the intermediate levels are the hashes of their immediate children. The root of the MHT is unique since the collision resistance property of hash function ensures that no two hash values differing by atleast 1 bit should be same. To adapt to an IoT environment, the traditional definition of MHT concepts have been modified.

Definition 1 Merkle Hash Assignment. *Let T be a MHT created in an IoT setup with n devices, i.e., having* $\log_2 n + 1$ *levels and let* $RFID_i$ *be the RFID of* i^{th} *IoT device in T. The Merkle hash assignment associated with the device with* $RFID_i$ *of T at level 1, denoted as* $\phi_{i,1}$ *is computed as*

$$\phi_{i,1} = H(RFID_i) \tag{1}$$

Similarly, the hash values of all node i except for the leaf nodes at level j denoted as $\phi_{i,j}$ *is computed by the following function:*

$$\phi_{i,j} = H(\phi_{2i-1,j-1} || \phi_{2i,j-1}) \tag{2}$$

where '||' denotes the concatenation operator and H(.) is the hash function.

According to Definition 1, the Merkle hash value associated with a device D_i is the result of a hash function applied to its RFID tag. To ensure the integrity of the Merkle hash value of the Trusted Centre (TC), we assume that the root value ϕ_{root} is tamper-resistant whose credentials have been thoroughly checked by top-level security system. Each leaf node in the constructed MHT can be verified through its Merkle Hash Path θ which is defined next.

Definition 2 Merkle Hash Path. *For each level* $l < \log_2 n + 1$ *(height of the tree), we define Merkle Hash Path* θ *to be the* ϕ *values of all the sibling nodes at each level l on the path connecting the leaf to the root node. The Merkle Hash Path signifying the authentication data at level l is then the set* $\theta_l | 1 \leq l \leq \log_2 n + 1$.

The authentication procedure of a leaf node is then carried out as: The ϕ value at the leaf is first hashed with its sibling θ_1, which, in turn, is hashed together with θ_2, and so on till the root is reached. At this stage, the calculated root value accumulated through the Merkle Hash Path is compared with equal to the known root value ϕ_root. If it turns out to be equal, then the leaf node is accepted as authentic. It is obvious that consecutive leaf nodes share a large portion of the authentication data θ when the leaves are ordered from left to right in the tree, thereby saving a lot of communication overhead in sending redundant data.

4   Modified Sinusoidal Quadratic Map

Chaotic maps are defined as mathematical functions that characterises the chaotic behaviour of the system and which depends on their initial conditions and control parameters. Our proposed chaotic map inspired by the classical quadratic map [36] is given as

$$x_{n+1} = 1 - \sin(r + ax_n^2) \text{ for } a > 3 \tag{3}$$

where the initial condition x_0, a and r are the control parameters. For our proposed equation, the value of a must be above 3 i.e., $a > 3$ in order to be chaotic.

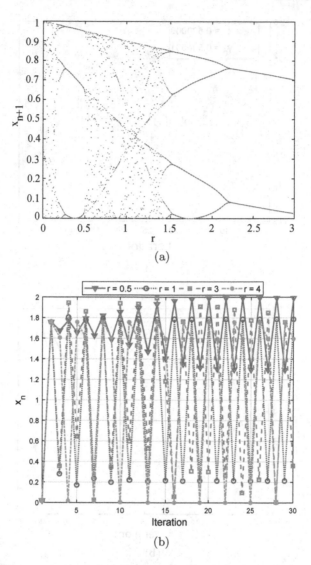

Fig. 1. (a) The proposed sinusoidal chaotic map for a = 4 (b) Convergence and period doubling plot for a = 3 and $x_0 = 0.02$

4.1   Analysis of the Proposed Map

Bifurcation plot in a dynamical system is a visual representation defining the behaviour of a system when its control parameters undergo some change [37]. The bifurcation diagram of the proposed map is shown in Fig. 1(a) where the solution was iterated for different values of r for a particular value of $a(a = 4)$. From the figure, the three regions of chaos, convergence and bifurcation can be observed clearly. The convergence region can be seen to start at $r = 2.2$, the

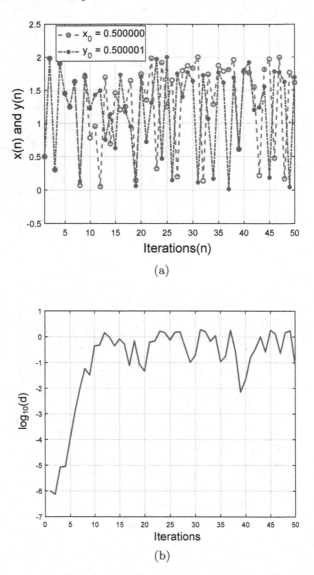

Fig. 2. (a) Sensitivity to initial conditions for two very close initial conditions of $x_0 = 0.500000$ and $x_0 = 0.500001$ for a = 4 and r = 3.9 (b) Semi-log plot for conditions of $x_0 = 0.500000$ and $x_0 = 0.500001$ for a = 4 and r = 3.9 (Color figure online)

bifurcation region lies in the range $r \in [1.5, 2.2]$ while the region $r \in [0, 1.5]$ can be considered to be chaotic with slight windows of stability that occurs in $r \in [0.3, 0.5]$. Each of these points are plotted for the particular value that the system settles towards for a specific value of r. Figure 1(b), on the other hand shows the phenomenon of period doubling where it can be seen that for $r = 0.5$, the system just starts becoming unstable. This is can correlated with the status

---

Algorithm 1. Authentication Algorithm using MHT

Input: Number of devices in the network n
Output: Authenticated Result

```
/* Construction of Merkle tree                                    */
```
1   Number of levels $l \leftarrow \log_2 n + 1$
2   **foreach** *level* $j \in l$ **do**
3 | **foreach** *device* $i \in N$ **do**
4 | | **if** $\phi_{i,j}$ *is a leaf node* **then**
5 | | | Calculate $\phi_{i,j} \leftarrow H(RFID_i)$
6 | | **else**
7 | | | $\phi_{i,j} \leftarrow H(\phi_{2i-1,j-1} || \phi_{2i,j-1})$

```
/* Authenticating an IoT device with RFID X                       */
```
8   **if** $\phi_{i,j} == \phi_{root}$ **then**
9 | X is authenticated to be a valid device

10 **else**
11 | A threat is detected

---

of the status at r = 0.5 in Fig. 1(a) where the diagram shows moderate chaos. Similarly, increasing the values of r results in increase in randomicity until r reaches 3 where the system fluctuates between two values that are the period attractors.

Sensitivity to initial conditions is another prime characteristic of chaos which states that two very close values of initial conditions diverge significantly over time. This phenomenon can be observed in Fig. 2(a) where two very close initial conditions $x_0 = 0.500000$ (in pink) and $y_0 0.500001$ (in blue) are iterated in the chaotic regime. As can be observed from the figure, the two trajectories are almost identical for the first 9 iterations. However, after the 10th iteration, the minuscule difference in the initial conditions diverges exponentially and show little in common as the number of iterations increases. This phenomenon is known as sensitivity to initial conditions and can be observed in our proposed chaotic map thus confirming its ramdomicity. Furthermore, a semi-log plot is constructed and presented in Fig. 2(b) that highlights the difference between the changes in the two initial conditions x_0 and y_0 over time. The difference d_n is calculated as $d_n = | x_n - y_n |$ and its logarithm values and plotted against the number of iterations n in Fig. 2(b). The exponential increase in the value $\log_{10} d$ can be clearly observed from the figure over the passage of time. This plot again gives an indication of the random nature of our map as the number of iterations increases which has been effectively exploited in our work for designing the encryption algorithm.

5 Proposed Security Protocol

Since our proposed architecture relates to an IoT scenario, adapting the known concepts of MHT and chaos theory to an IoT environment was necessary. The processes are explained in detail next.

5.1 Registration

Registration is the first phase where all the n IoT devices $D_i(i = 1, \ldots n)$ wishing to form a network register themselves with the Trusted centre (TC) with their designated n unique RFID tags. TC then constructs the tree by deriving the hash value of each $RFID_i$ and stores them in a table for future authentication. TC then distributes the hash values to all leaf nodes. Thus, only the authenticated devices in the network are in possession of the hash values that is required for their transmission of data.

5.2 Authenticating the Devices Using MHT

As mentioned earlier, authentication is performed by TC which is assumed to be secured from any form of attacks and whose credentials are verified from the top-level security system. Any device D_i wishing to initiate data transfer has to first test for its authenticity. This is done by sending a request message REQ_i to TC that indicates its desire for communication. TC, on receiving REQ_i, asks for the proof that D_i belongs to the network and is a valid device. D_i now sends hash values of the merkle hash path (θ) for authentication. On receiving the proofs, TC calculates the hash value using Eq. 2, and checks if its stored hash of the root (i.e., h_{root}) is equal to the calculated hash. If the two hashes match, the device D_i is an authenticated device and can proceed to the process of data exchange. Algorithm 1 illustrates the process of authentication for our proposed system where each device presented as a leaf node is authenticated by recursively computing and concatenating the hash values along the Merkle Hash path. Since only the hash functions are computed, the computation cost of verification is very low.

5.3 Establishment of Keys

In our work, the key of encryption algorithm is produced by our proposed chaotic map because of its marked nature of randomness. It is a known fact that the more random a key is, the more difficult it is for an attacker to break. Therefore on the basis of the bifurcation diagram, it is easy to note the areas the map produces random chaotic behaviour. The control parameters needed to generate the bifurcation diagram comes from the MHT. Our proposed chaotic map takes as input three control parameters: a pre-shared session key $\mathcal{S}$, a key $\mathcal{K}$ which acts as the initial condition of the map and the number of iterations Itr that signifies the number of times $\mathcal{K}$ is iterated in the map. Based on the number of levels l, keys are produced. These keys are nothing but the value of ϕ at each node in the merkle hash path θ for the device D_i. After which, $\mathcal{K}$ is calculated as follows

$$\mathcal{K} = key_1 \oplus key_2 \oplus key_3 \oplus ... \oplus key_l$$

The value of $\mathcal{K}$ in binary is converted into decimal that serves as the initial condition in the chaotic map. The pre-shared session key $\mathcal{S}$ is generated which is a random number such that the value > 3. This is set keeping in mind that our proposed chaotic map is chaotic in this range. The iteration number Itr is calculated using both the values of $\mathcal{K}$ and $\mathcal{S}$. The number of digits in $\mathcal{K}$, say, dig is estimated. Now, in the generated value of $\mathcal{S}$, dig digits after decimal is extracted and summed up with the value of $\mathcal{K}$ that yields the iteration number.

Theorem 1. *The key space size of our proposed encryption algorithm is $2^{280} \times l$ where l denotes the number of levels in the Merkle Hash Tree.*

Proof. Since the hash function used in our protocol is SHA-1, which produces a 160-bit binary output, the key space required for $\mathcal{K}$ alone is 2^{160}. Furthermore, in order to ensure the dynamical system falls in the chaotic regime, the range of a that is also the pre-shared session key is restricted to 32-digits values for $a > 3$. This value a or $\mathcal{S}$ is a 32 bit decimal number that is generated randomly. Since the ASCII table supported by MATLAB is composed of 128 values. Each of these 40-hex digits (output of SHA-1) are mapped to its corresponding ASCII, the maximum value of which is 128. Therefore, the maximum value of the hash value at each node, i.e., key_1, key_2,...,key_l (l is the number of levels), cannot exceed $40 \times 128 = 5120$ which requires $\approx$ 13-bits each to represent. Therefore, the value of $\mathcal{K}$ which is the XOR operation of $key_1, key_2, ..., key_l$ also comprises of 13-bits. Hence, the iteration number Itr that is dependent on the number of digits in $\mathcal{K}$ should also not exceed $13 + 1$ (for carry) bits. Summing it all up, the key size for our proposed algorithm is $(2^{160} \times l) \times 10^{32} \times 2^{13} \approx 2^{280} \times l$, where l is the number of levels in the Merkle Hash Tree (Fig. 3).

5.4 Data Encryption/Decryption

The value of ϑ obtained after the iteration process of the chaotic map is then combined with the plain text $\mathcal{P}$ using a XOR operation along with the previous ciphertext value. The inclusion of previous ciphertext for XOR operation was adopted for ensuring dynamic feedback in our proposed architecture. Thus, at any instant t the encrypted data will be given by $Cipher = \mathcal{P}^t \oplus \vartheta \oplus Cipher^{t-1}$. Algorithm 2 displays the essential steps of our chaos based encryption/decryption algorithm. This function takes as an input a 256-bits plaintext $\mathcal{P}^t$ data. In order to add provision for integrity check, an alternative coding approach that appends a count of the '0' bits $N^{(0)}$ in C^t before communicating it to TC. The new message, $\mathcal{C}$, would be only be $\log_2 256\text{-bit} = 8$ bits longer than the original 256-bit message, $Cipher^t$. After appending the zero count $N^{(0)}$ the final ciphertext $\mathcal{C}$ is

---

Algorithm 2. Proposed Chaotic Encryption algorithm

Input: Raw data $\mathcal{P}$

Output: Encrypted data $\mathcal{C}$

1 Generate the keys $key_1, key_2, .., key_l$ based on the number of levels l
2 Final key $\mathcal{K} \leftarrow key_1 \oplus key_2 \oplus, ...key_l$
3 Convert $\mathcal{K}$ into binary
4 $dig \leftarrow$ number of decimal digits of $\mathcal{K}$
5 $\delta \leftarrow$ take dig digits after decimal from pre-shared session key $\mathcal{S}$
6 Iteration $Itr \leftarrow \mathcal{K} + \delta$
   /* Set initial condition $\mathcal{K}$ and $\mathcal{S}$ as control parameter and iterate
      in the chaotic map Itr number of times   */
7 ChaosVal $\leftarrow$ Chaos$(\mathcal{K}, \mathcal{S}, Itr)$
8 $\vartheta \leftarrow$ ChaosVal $\times Itr$
9 **if** $\mathcal{P}$ *is the first plaintext after registration* **then** $Cipher^t \leftarrow \mathcal{P}^t \oplus \vartheta$
10 **else** $Cipher^t \leftarrow \mathcal{P}^t \oplus \vartheta \oplus Cipher^{t-1}$
11 $\mathcal{C} \leftarrow \mathcal{C}^t | \mathcal{N}^{(0)}$
12 **return** $\mathcal{C}$

---

sent to TC for data fusion through the communication medium. TC, on receiving $\mathcal{C}$, first checks whether the message has been tampered with by comparing the last 8 bits that signifies $\mathcal{N}^{(0)}$ with the number of zeros in the first 256-bits in $\mathcal{C}$. If the values do not match, a security threat is detected and subsequent actions are undertaken to remedy the problem. If however, the values match, the ciphertext is assumed to be free of any tampering by an intruder and thus is further processed to extract the plaintext. In our work, since the merkle hash values ϕ is used for modulation in the chaotic map, which is known to both D_i and TC, both can generate the chaotic initial value $\mathcal{K}$ and the Itr value individually. In the decryption process, utilizing the symmetric property of XOR operation TC decrypt the received data $Cipher^t$ as $Cipher^t \oplus \vartheta \oplus Cipher^{t-1} = (\mathcal{P}^t \oplus \vartheta \oplus Cipher^{t-1}) \oplus \vartheta \oplus Cipher^{t-1}$ which equals $\mathcal{P}^t$.

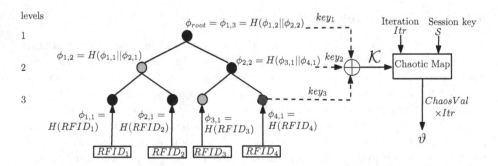

Fig. 3. Key generation using Merkle Hash Tree together with the proposed chaotic map

5.5   Secure Data Fusion

Since the main aim of our work is security in the process of data fusion. Therefore the authentication, key establishment, encryption processes are extended to all the IoT devices in the network that participate in the data fusion process. TC is the centre for data fusion, where encrypted data from all the devices arrive. Based on which the process of fusion takes place to arrive to a conclusion. After which, the decision is communicated to the concerned authority. For instance, in real-time monitoring environment, any critical event needs to be conveyed at real time. The assessment of the critical event is done by fusion of the data appropriately at any time instant at the same time ensuring its security in the process. This idea can be described as follows: assume P_i^t, C_i^t represent t^{th} plaintext and ciphertext respectively from the ith device in the network. Since, a major portion of an device's energy is consumed during the process of communication, reducing the data transfer in the network is useful in saving battery life of energy-constraint IoT devices. To minimize the number of transmissions from thousands of devices towards the TC, a single session key $\mathcal{S}$ is used for all devices for a particular session; after which it becomes obsolete. This limits the number of transmission in the network as well as ensures the security since $\mathcal{S}$ is not the only control parameter that is needed to construct the chaotic map. Thus, TC, on receiving the encrypted message from n devices, decrypts it and performs its computations to derive its result. The data fusion algorithm used by TC is beyond the scope of this paper and thus is avoided to facilitate easy understanding.

6   Experiment

For the sake of simplicity in our experiment, we have simulated an IoT environment consisting of 8 IoT devices $D_i | i = 1, 2, .., 8$ in MATLAB, each generating time varying data P^t every t time instant. For instance, sensory data (referred to as the plaintext) generated by device D_2 is given as $\mathcal{P}_2 = P_2^1, P_2^2, \ldots, P_2^{t-1}, P_2^t$. For our proposed chaotic map $x_{n+1} = 1 - sin(r + ax_n^2)$ for $a \in \{1, 4\}$, the control parameters are the pre-shared session key $\mathcal{S}$, the initial condition x_0 denoted as $\mathcal{K}$ and the iteration number Itr.

Registration

- Step 1: All the 8 IoT devices in the network register themselves with the trusted data fusion centre (TC) with their designated RFID tags, $RFID_i | i = 1, .., 8$. RFIDs are 96-bit binary numbers or 24 hex digits. For our experiment, we have used random 24 hex numbers as RFIDs as shown in Table 2.
- Step 2: TC creates a Merkle Hash Tree with all the devices in the network, in which all the leaf nodes the RFIDs of the devices. The tree is constructed as shown in Fig. 4.

Table 2. RFID tags corresponding to each device for our experiment

Device	RFID tags
D_1	45 3d 6c e1 48 16 85 57 e3 29 c5 89
D_2	77 c0 23 0e b5 0e 39 63 3a 48 5b bf
D_3	2e e0 62 6d 14 ca e6 83 18 7a e7 9e
D_4	ba 9d 08 f4 2b 4b 5e 23 51 d5 70 2a
D_5	5a 3e ed 7e b4 7f d3 e8 60 40 77 37
D_6	3b 0a f1 4a 7c e9 14 ca ac da 3f c9
D_7	21 f5 cb a5 80 0f 82 58 aa 90 f2 d5
D_8	b4 e0 72 d6 50 72 3a cd 67 85 5c ab

– Step 3: In addition, TC maintains a table where RFID tag of each device is stored. A randomly generated pre-shared session key $\mathcal{S} \in \{1, 4\}$ is stored for each time instant t. This session key is used by all the devices for communication for each session, at the expiration of which, the session key $\mathcal{S}$ becomes obsolete.

Authentication using MHT

– Step 4: Device D_8 deciding to initiate a communication does so by sending a request message REQ_8 to TC. Figure 4 depicts this situation where device D_8 is denoted by a red circle.
– Step 5: TC on receiving REQ_8, asks for the proof that D_8 belongs to the network and is a valid device.
– Step 6: D_8 now sends hash values of the merkle hash path for authentication. That is, D_8 sends the value of $\phi_{8,1}, \phi_{7,1}, \phi_{3,2}$ and $\phi_{1,3}$ as authentication proofs to the TC. The proofs/sibling nodes are highlighted in blue and the initiator node $\phi_{8,1}$ in red in Fig. 4.
– Step 7: TC, on receiving the proofs calculates the resultant hash value from the individual hash values received from D_8 according to Eq. 2, as

$$
\begin{aligned}
\phi_{1,4} &= H(\phi_{1,3} || \phi_{2,3}) \\
&= H(\phi_{1,3} || H(\phi_{3,2} || \phi_{4,2})) \\
&= H(\phi_{1,3} || H(\phi_{3,2} || H(\phi_{7,1} || \phi_{8,1})))
\end{aligned}
$$

– Step 8: TC now compares the resultant value $\phi_{1,4}$ with its own hash ϕ_{root}. The $\phi_{1,4}$ obtained in the previous step matches with the stored value of ϕ_{root} in our experiment. Since the two hashes are equal, the device D_8 is an authenticated device and thus can proceed to the process of data exchange.

Key Generation and Exchange

– Step 9: Based on the number of levels in the Merkle hash tree, a unique key is generated at each level denoted as $key_1, key_2, ..., key_n$ which is nothing but

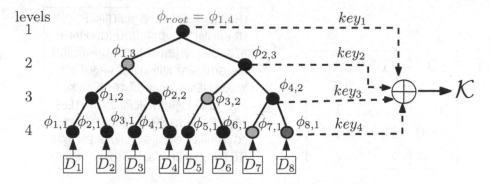

Fig. 4. Key generation using Merkle Hash Tree together with the proposed chaotic map (Color figure online)

the hash values of each node in the Merkle Hash Path. In our experiment number of levels is 4, therefore four keys are generated.

– Step 10: The values of keys i.e., key_1, key_2, key_3 and key_4 are not the proofs that are communicated in the authentication process, but rather the ϕ values of those nodes that fall in the path from the device to the TC. That is, for the device D_8, the keys as illustrated in the the Fig. 4 are $\phi_{8,1}$, $\phi_{4,2}$, $\phi_{2,3}$ and $\phi_{1,4}$ after converting them to their ASCII and summing the resultant values.

– Step 11: All 4 keys key_1, key_2, key_3 and key_4 are XORed to produce the final key $\mathcal{K}$, after which it is converted into decimal that serves as the initial condition for our chaotic map. In our experiment, value of $\mathcal{K}$ (note that the XOR operations are all done in binary, in order to save space, the values are replaced by their decimal equivalent) is

$$\mathcal{K} = 2787 \oplus 2736 \oplus 2671 \oplus 2907$$
$$= 2620$$

The value of $\mathcal{K} = 2620$ is the initial condition for the map generation.

– Step 12: At this point, a shared session key $\mathcal{S}$ is generated between the TC and D_8. $\mathcal{S} \in \{1, 4\}$ is generated such that it falls in the chaotic region of the proposed map. For our experiment we have randomly generated the value as

$$\mathcal{S} = 3.0362054645733205227031703543616$$

– Step 13: The value of Itr that denotes the number of iterations is calculated next. The device/TC first estimates the number of digits in $\mathcal{K}$ as dig. In our experiment, $dig = 4$. Adding dig number of digits after decimal of the value $\mathcal{S}$ to the value of $\mathcal{K}$ yields the value of Itr. That is,

$$Itr = 4 \text{ (value of dig) digits after decimal of } \mathcal{S} + \mathcal{K}$$
$$= 0362 + 2620$$
$$= 2982$$

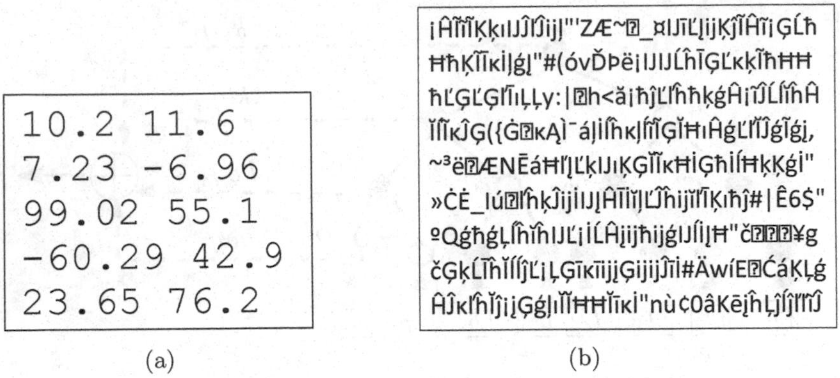

```
10.2  11.6
7.23  -6.96
99.02  55.1
-60.29  42.9
23.65  76.2
```

(a)

¡ÂꭲꭧκκιⅠJⅠ⍳ij|'''ZÆ~▨_ꭤιJⅰ⌊ij⊼ĵⅠÂꭲⅰ⌷∫Ⱡꞛ
Ⱨꞛ⊼ⅡꞚⅰ|Ɫ|"#(óvⅮⱣёⅰⅠJⅠJⱠⱨⅠⱤꞛⱠκⱪꞛꞛⱧⱧ
ꞕⱢꞛⱢⱤⱣⅡⱢⱢꞙ:|⊞h<ㄡⅰⱧĵⱢꞕⱧⱨꞛⱥÂⅰꭲⱭⱢꞕⱧÂ
ⱥꞕⱪⅉⱤ({ⱤⅇκⱭⱦ⁻ㄠ|ⱥꞕⱨκ|ꭲꞕⱤꞕⱧⅰÂⱥⱢⱭⱭⱤꞕㄱㄠ,
~³ⱸⱥⱮꞒ⍤ƝꞒㄠⱧⱤⱤ'ⱢⱪⱭJⱭꞍⱤꞕⱨꞕⱪⱧⅰⱷⱨⱪⱪⱥㄠ"
»ⱢꞒ_⍳úⅇⱭꞕⱨⱪⱭⱭⱭⅠⅠJ|ⱢⱯꞕⅰ|ⱢꞕⱧⱭⱮꞕꞕⱪⱭⱨⱥ#|Ê6$"
ºⱢⱥꞕⱥⱢⱬꞕⱧꞕⱧⅠJⱢꞕⅰⱢꞕⱯⱭⱭⱥꞕⱧꞕⱮⱥⱢⱭꞕ⌷Ɫ"㉠⊡⊡⊡¥g
ㄩⱤκⱢⅠⱧꞕⱢ'ꞕⱢ'ⱭⱢⅠⱤⅰκⅰⅰⱭⱭꞕⱭⱭⱭⱭⱥꞕ#Ⱥ⍹ⅰꞒⅇⱢⱭ⌀ⱩⱥⱢⱤ
ÂⱭκⱭⱨꞕⱭ⌀⌷ⱤⱤⱪⱭꞕⅡⱢⱧⱧ⍹Ɫκⅰ"nùⱢⱭㄠⱪꞓⱣ|ⱨⱢⱭ|ⱢⱭⱭⱭⱭⱢꞕⱭ

(b)

Fig. 5. Analysis of our proposed Chaotic map for (a) Plaintext (b) Encryption with correct key

Encryption

– Step 14: The values of $ChaosVal$ and ϑ after the map is iterated Itr number of times are as follows:

$$ChaosVal = -1.2054$$
$$\vartheta = ChaosVal \times Itr$$
$$= -1.2054 \times 2982 = -3594.4$$

– Step 15: The plaintext produced by D_8 at time instant t and (t−1) in our experiment be given as $\mathcal{P}^{t-1} = 10.2$ and $\mathcal{P}^t = 11.6$. The plaintext is encrypted as explained in Sect. 5.4.

– Step 16: The next step is to perform XOR operation of the plaintext $\mathcal{P}^t$, chaotic output ϑ and the previous cipher $\mathcal{C}^{t-1}$ after converting them into their binary 256-bit equivalent as

$$\mathcal{C}^t = \begin{cases} \mathcal{C}^t = \mathcal{P}^t \oplus \vartheta \oplus \mathcal{C}^{t-1} & \text{if } t \neq 1 \text{ or,} \\ \mathcal{P}^t \oplus \vartheta & \text{otherwise} \end{cases}$$

– Step 17: Similarly, for integrity checking since the size of $\mathcal{C}^t$ is 256 bits, then the maximum size of N^0 will be $\log_2 256 = 8$ bits.

– Step 18: The resultant $\mathcal{C}^t$ i.e., $256 + 8$ bits is converted into their ASCII values generating $32 + 1$ characters of ASCII which is the final cipher $Cipher$. Figure 5(a) and (b) shows the plaintext/sensor data corresponding to each of the 8 ciphertexts respectively.

– Step 19: The ciphertext $Cipher$ is then sent to TC for decrypting.

Decryption

– Step 20: TC, on receiving the ciphertext $Cipher$, now performs the reverse operation by first estimating the value $\mathcal{K}$ from the information provided by the device D_8.

- Step 21: Value of Iteration Itr and is calculated using similar approach with the help of the pre-shared session key S. Ultimately ϑ is calculated through iteration on the chaotic map with the help of the control parameters, i.e., K, S and Itr.
- Step 22: TC extracts the plaintext by performing XOR operation of the previous known cipher C_{i-1}, the current cipher value C_i and the output produced by the chaotic map ϑ.
- Step 23: The value obtained after the XOR operation is the plaintext P after converting to its decimal form, i.e., the value of 10.2 is successfully decrypted by the TC.

Secure Data Fusion

- Step 24: TC individually decrypts the values from each of the devices. After the Ciphertexts $C_1^t, C_2^t, \ldots, C_8^t$ from devices $D_1, D_2, \ldots, D_8$ respectively are successfully decrypted into the plaintexts $P_1^t, P_2^t, \ldots, P_8^t$, the process of data fusion begins.
- Step 25: In this phase, all the sensor information in the form of plaintexts are fused to form a decision. (Note that the algorithm for data fusion is beyond the scope of this paper and thus is not added in order to avoid complication.)
- Step 26: The decision is conveyed to the concerned authority to the IoT application wirelessly or through the monitoring app.

All the steps above are explained with respect to a single device D_8. Similar procedure is followed by all devices in the network, i.e., $D_i | i = 1 \ldots 8$ each follow through the steps of authentication, key exchange, encryption and decryption.

7   Security Analysis

Accomplishment of Anonymity. Anonymity ensures that even if the attacker A eavesdrops on any ongoing communication, he/she should not be able to detect the identity of either the sender or receiver of the intercepted message, i.e. the identity of a device D_i is completely anonymous. This is achieved in our protocol by the one-way property of Hash functions which is the heart of our work in this paper. Intuitively, a one way function is one which is easy to compute but difficult to invert. Thus, even if the hash proofs of the device D_i are intercepted by the attacker, it is impossible for him/her to extract the identity or the RFID of D_i thus achieving device anonymity.

Accomplishment of the Device Authentication. Since our security protocol is based on the MHT where the values at each node are the hash of RFID tags and the RFID tags uniquely identifies a device, the generated hash values are also unique. In our protocol, any attacker attempting to initiate communication with the TC cannot forge the RFID of the authentic device and thus cannot deliver the accurate proofs. Moreover, even if the attacker in some way intercepts the RFID of the device, it is impossible to get hold of the hash values of all its siblings that constitutes the valid proof. In this way, TC authenticates a legitimate device and prevents unwanted communication from untrusted third parties.

Accomplishment of Data Integrity. To achieve data integrity, we have incorporated a mechanism where the ciphertext includes few bits for integrity checking. Suppose the ciphertext C be 11000110. The number of 0s is 4 or $N^{(0)} = 100$. Then $Cipher$ would be $Cipher = 11000110|100$ (where | signifies the division of $Cipher$ into C^t and its 0 count $N^{(0)}$). Now if the ciphertext 11000110 were tampered by an attacker to 11000100 by changing the seventh bit to a 0, the value of $N^{(0)}$ being the same, the cipher would then be $Cipher = 11000100|100$. For the $Cipher$ to be a valid codeword, the count $N^{(0)} = 100$ would also have to be changed to 101 because we now have 5 0s, not 4. But this requires changing a 0 to a 1, something that is forbidden. If the codeword were changed to 11000110|110 by altering $N^{(0)}$, then C would have to be changed so that it had 6 0's instead of 4. Again, this requires changing a 0 to 1 which is not possible. In this way, our algorithm guarantees data integrity.

Accomplishment of the Data Confidentiality. Data confidentiality is maintained in our protocol as there is no requirement of exchanging keys in the network. As a result, any attempt by an unauthorized user to forge identity is nullified. Therefore, the entire process of our proposed architecture is highly confidential and the exchanged data is highly secured against tampering.

Resistance to Replay Attacks. Our proposed is resilient to replay attacks by using a random value of S. In this way, an attacker cannot replay the same message again and again with the intention of passing the authentication phase. Others key parameters such as K and Itr are not shared in the insecure channel and thus they cannot be intercepted by the attacker. These keys are generated at both ends separately through the control parameters, thus making our proposed scheme secure against from unwanted replay attacks.

Resistance to Forgery Attacks. An attacker may also attempt to use the RFID of any legal validated device to pass the verification process of the TC. In that case, the attacker needs to construct a valid request message REQ with valid proofs to pass the TC's verification. However, to do that, he/she needs to not only know the hash of the RFID but in addition the individual hash values of all the sibling nodes in its path to the TC i.e. apart from $H(RFID)$, other proofs that includes the ϕ values calculated for every node j at level i as $\phi_{i,j} = H(\phi_{2i-1,j-1}||\phi_{2i,j-1})$, which is quite impossible for him/her to figure out as these are the unknown secrets and therefore an attacker cannot convince TC of its identity. In this way, our proposed scheme can resist forgery attacks.

8 Conclusion

Owing to the urgent need for developing security algorithms for Internet of Things (IoT) environment, this paper presents a security protocol by combining the advantages of both Merkle Hash Tree and Chaotic Cryptography. Our contribution is two-fold. First, we develop an authentication protocol based on Merkle Hash Tree that we have improved to suit to an IoT application by utilizing the RFID tags for generating the tree. Secondly, we have designed an encryption algorithm inspired by the chaos theory in cryptography. Additionally, we have proposed a novel chaotic map that has been used for designing the encryption algorithm. The proposed security protocol use lightweight computations that is well suited for the resource-constrained IoT devices. Experimental and security analysis proves the effectiveness of our algorithms and its resilience to security attacks.

Acknowledgement. This work is supported by the Ministry of Electronics & Information Technology (MeitY), Government of India under the Visvesvaraya PhD Scheme for Electronics & IT (PhD-PLA/4(71)/2015-16).

References

1. Guo, B., Zhang, D., Yu, Z., Liang, Y., Wang, Z., Zhou, X.: From the Internet of Things to embedded intelligence. World Wide Web **16**(4), 399–420 (2013)
2. Satyadevan, S., Kalarickal, B.S., Jinesh, M.K.: Security, trust and implementation limitations of prominent IoT platforms. In: Satapathy, S.C., Biswal, B.N., Udgata, S.K., Mandal, J.K. (eds.) Proceedings of the 3rd International Conference on Frontiers of Intelligent Computing: Theory and Applications (FICTA) 2014. AISC, vol. 328, pp. 85–95. Springer, Cham (2015). https://doi.org/10.1007/978-3-319-12012-6_10
3. Weber, R.H.: Internet of Things-new security and privacy challenges. Comput. Law Secur. Rev. **26**(1), 23–30 (2010)
4. Lampropoulos, K., Denazis, S.: Identity management directions in future Internet. IEEE Commun. Mag. **49**(12), 74–83 (2011)
5. Suhardi, R.A.: A survey of security aspects for Internet of Things in healthcare. In: Kim, K., Joukov, N. (eds.) Information Science and Applications (ICISA) 2016. Lecture Notes in Electrical Engineering, vol. 376. Springer, Singapore (2016). https://doi.org/10.1007/978-981-10-0557-2_117
6. Alasmari, S., Anwar, M.: Security & privacy challenges in IoT-based health cloud. In: 2016 International Conference on Computational Science and Computational Intelligence (CSCI), pp. 198–201. IEEE (2016)
7. Islam, S.R., Kwak, D., Kabir, M.H., Hossain, M., Kwak, K.-S.: The Internet of Things for health care: a comprehensive survey. IEEE Access **3**, 678–708 (2015)
8. Roman, R., Najera, P., Lopez, J.: Securing the Internet of Things. Computer **44**(9), 51–58 (2011)
9. Kalra, S., Sood, S.K.: Secure authentication scheme for IoT and cloud servers. Pervasive Mob. Comput. **24**, 210–223 (2015)
10. Amin, R., Kumar, N., Biswas, G., Iqbal, R., Chang, V.: A light weight authentication protocol for IoT-enabled devices in distributed cloud computing environment. Future Gener. Comput. Syst. **78**, 1005–1019 (2018)

11. Zhou, L., Li, X., Yeh, K.-H., Su, C., Chiu, W.: Lightweight IoT-based authentication scheme in cloud computing circumstance. Future Gener. Comput. Syst. **91**, 244–251 (2019)
12. Mookherji, S., Sankaranarayanan, S.: Traffic data classification for security in IoT-based road signaling system. In: Nayak, J., Abraham, A., Krishna, B.M., Chandra Sekhar, G.T., Das, A.K. (eds.) Soft Computing in Data Analytics. AISC, vol. 758, pp. 589–599. Springer, Singapore (2019). https://doi.org/10.1007/978-981-13-0514-6_57
13. Nesa, N., Ghosh, T., Banerjee, I.: Outlier detection in sensed data using statistical learning models for IoT. In: 2018 IEEE Wireless Communications and Networking Conference (WCNC), pp. 1–6. IEEE (2018)
14. Nesa, N., Ghosh, T., Banerjee, I.: Non-parametric sequence-based learning approach for outlier detection in IoT. Future Gener. Comput. Syst. **82**, 412–421 (2018)
15. Rathore, H., Jha, S.: Bio-inspired machine learning based wireless sensor network security. In: 2013 World Congress on Nature and Biologically Inspired Computing, pp.140–146. IEEE (2013)
16. Bodei, C., Chessa, S., Galletta, L.: Measuring security in IoT communications. Theor. Comput. Sci. **764**, 100–124 (2019)
17. Bodei, C., Degano, P., Ferrari, G.-L., Galletta, L.: Where do your iot ingredients come from? In: Lluch Lafuente, A., Proença, J. (eds.) COORDINATION 2016. LNCS, vol. 9686, pp. 35–50. Springer, Cham (2016). https://doi.org/10.1007/978-3-319-39519-7_3
18. Nesa, N., Ghosh, T., Banerjee, I.: Design of a chaos-based encryption scheme for sensor data using a novel logarithmic chaotic map. J. Inf. Secur. Appl. **47**, 320–328 (2019)
19. Shukla, P.K., Khare, A., Rizvi, M.A., Stalin, S., Kumar, S.: Applied cryptography using chaos function for fast digital logic-based systems in ubiquitous computing. Entropy **17**(3), 1387–1410 (2015)
20. Wang, W., et al.: An encryption algorithm based on combined chaos in body area networks (2017). http://www.sciencedirect.com/science/article/pii/S0045790617324138
21. Hamad, N., Rahman, M., Islam, S.: Novel remote authentication protocol using heart-signals with chaos cryptography, In: International Conference on Informatics, Health & Technology (ICIHT), pp. 1–7. IEEE (2017)
22. Ning, H., Liu, H., Yang, L.T.: Aggregated-proof based hierarchical authentication scheme for the Internet of Things. IEEE Trans. Parallel Distrib. Syst. **26**(3), 657–667 (2015)
23. Liu, J., Su, H., Ma, Y., Wang, G., Wang, Y., Zhang, K.: Chaos characteristics and least squares support vector machines based online pipeline small leakages detection. Chaos, Solitons Fractals **91**, 656–669 (2016)
24. Furquim, G., Pessin, G., Faiçal, B.S., Mendiondo, E.M., Ueyama, J.: Improving the accuracy of a flood forecasting model by means of machine learning and chaos theory. Neural Comput. Appl. **27**(5), 1129–1141 (2016)
25. Furquim, G., Mello, R., Pessin, G., Faiçal, B.S., Mendiondo, E.M., Ueyama, J.: An accurate flood forecasting model using wireless sensor networks and chaos theory: a case study with real WSN deployment in Brazil. In: Mladenov, V., Jayne, C., Iliadis, L. (eds.) EANN 2014. CCIS, vol. 459, pp. 92–102. Springer, Cham (2014). https://doi.org/10.1007/978-3-319-11071-4_9
26. Yang, L., Fei, L.Y., Dong, Y.X., Yan, H.: Iris recognition system based on chaos encryption. In: 2010 International Conference on Computer Design and Applications (ICCDA), vol. 1, pp. V1–537. IEEE (2010)

27. Liu, X., Fang, X., Qin, Z., Ye, C., Xie, M.: A short-term forecasting algorithm for network traffic based on chaos theory and SVM. J. Netw. Syst. Manage. **19**(4), 427–447 (2011)
28. Li, H., Lu, R., Zhou, L., Yang, B., Shen, X.: An efficient merkle-tree-based authentication scheme for smart grid. IEEE Syst. J. **8**(2), 655–663 (2014)
29. Li, D., Aung, Z., Williams, J.R., Sanchez, A.: Efficient authentication scheme for data aggregation in smart grid with fault tolerance and fault diagnosis. In: 2012 IEEE PES Innovative Smart Grid Technologies (ISGT), pp. 1–8. IEEE (2012)
30. Nicanfar, H., Jokar, P., Leung, V.C.: Smart grid authentication and key management for unicast and multicast communications. In: 2011 IEEE PES Innovative Smart Grid Technologies, pp. 1–8. IEEE (2011)
31. Xu, K., Ma, X., Liu, C.: A hash tree based authentication scheme in SIP applications. In: IEEE International Conference on Communications, 2008. ICC 2008, pp. 1510–1514. IEEE (2008)
32. Liu, C., Ranjan, R., Yang, C., Zhang, X., Wang, L., Chen, J.: MuR-DPA: top-down levelled multi-replica merkle hash tree based secure public auditing for dynamic big data storage on cloud. IEEE Trans. Comput. **64**(9), 2609–2622 (2015)
33. Zhang, H., Tu, T., et al.: Dynamic outsourced auditing services for cloud storage based on batch-leaves-authenticated Merkle hash tree. IEEE Trans. Serv. Comput. **PP**(99), 1 (2017)
34. Garg, N., Bawa, S.: RITS-MHT: relative indexed and time stamped Merkle hash tree based data auditing protocol for cloud computing. J. Netw. Comput. Appl. **84**(Supplement C), 1–13 (2017). http://www.sciencedirect.com/science/arti cle/pii/S1084804517300668
35. Merkle, R.C.: A certified digital signature. In: Brassard, G. (ed.) Advances in Cryptology — CRYPTO 1989 Proceedings. CRYPTO 1989. Lecture Notes in Computer Science, vol. 435, pp. 218–238. Springer, New York (1990). https://doi.org/10.1007/0-387-34805-0_21
36. Moreira, F.J.S.: Chaotic dynamics of quadratic maps. IMPA (1993)
37. Lawande, Q., Ivan, B., Dhodapkar, S.: Chaos based cryptography: a new approach to secure communications, vol. 258, no. 258. BARC newsletter (2005)

A Deployment Framework for Ensuring Business Compliance Using Goal Models

Novarun Deb[2(✉)], Mandira Roy[1], Surochita Pal[1], Ankita Bhaumick[1],
and Nabendu Chaki[1]

[1] University of Calcutta, Kolkata 700106, WB, India
roy.mandiracs@gmail.com, pal.surochita@gmail.com,
ankitabhaumik6@gmail.com, nabendu@ieee.org
[2] Università Ca' Foscari, Via Torino, 153, 30172 Venezia (VE), Italy
novarun.deb@unive.it

Abstract. Based on initial research to transform a sequence agnostic goal model into a finite state model (FSM) and then checking them against temporal properties (in CTL), researchers have come up with guidelines for generating compliant finite state models altogether. The proposed guidelines provide a formal approach to prune a non-compliant FSM (generated by the Semantic Implosion Algorithm) and generate FSM-alternatives that satisfy the given temporal property. This paper is an extension of the previous work that implements the proposed guidelines and builds a deployment interface called $i^*ToNuSMV$ *3.0*. The working of the framework is demonstrated with the help of some use cases. In the end, a comparative study of the performance between the previous and current versions of the Semantic Implosion Algorithm (SIA) with respect to the size of the solution space and the execution times, respectively, has also been presented.

1 Introduction

Goal Oriented Requirements Engineering (GORE) techniques help system developers to establish rationales for the system being developed. GORE frameworks (like i* [15], KAOS [11], etc.) helps enterprise architects to develop a deep understanding of the of the problem domain, interests, priorities and abilities of various stakeholders [1,15]. Stand-alone goal models are not sufficient in spite of all the various analyses that can be performed on them. Goal model compliance to business rules still remains a challenge as goal models are sequence agnostic whereas business rules are inherently specified as a temporal ordering of events. Such temporal properties and constraints cannot be verified against goal models in general.

There have been some significant research initiatives in this direction. Formal Tropos [6] introduces the notion of integrating formal methods with goal models for performing different kinds of formal analysis, such as consistency checking, animation of the specification, and property verification. This is done by combining the primitive concepts of i* with a rich temporal specification language

© Springer-Verlag GmbH Germany, part of Springer Nature 2020
M. L. Gavrilova et al. (Eds.): Trans. on Comput. Sci. XXXV, LNCS 11960, pp. 106–118, 2020.
https://doi.org/10.1007/978-3-662-61092-3_6

inspired from KAOS. The T-tool [6] has been developed, based on the NuSMV [5] model verifier, to support this framework. In [2], the authors provide an efficient algorithm to transform goal models into finite state models while significantly reducing the size of the solution space (as compared to the naïve approach proposed previously). The i*ToNuSMV 2.02 tool [3] performs this model transformation and then maps the finite state model into the NuSMV input language. Users can verify temporal properties and constraints (specified in CTL) using the NuSMV model verifier that has been integrated into the tool interface.

The i*ToNuSMV 2.02 tool does not ensure finite state models to be compliant with temporal properties. It can only check and verify whether a model satisfies or contradicts a specified property. Driven by the need to generate compliant finite state models, researchers have proposed some initial guidelines in [4]. The authors show with the help of use-cases how the finite state model (generated by i*ToNuSMV 2.02) can be pruned with respect to a given CTL property, in order to make them compliant. The pruning process is a one-to-many transformation as there may exist more than one way of making a finite state model compliant. Each possible alternative is derived by the proposed framework. In this paper, the work presented in [4] is extended and a tool interface is built that implements the research guidelines suggested in the previous work. The i*ToNuSMV 3.0 deployment framework is presented as an updated version of the existing i*ToNuSMV 2.02 tool that allows the generation of compliant finite state models. This requires a temporal property (in CTL) to be fed as input along with the goal model description. The output finite state models are compliant with this given CTL property and can be cross-checked with the in-built NuSMV model verifier. Some use cases are also presented to demonstrate how the framework functionalities. Finally, some experimental results are documented to show how the performance of the model checking process is improved by pruning the finite state models and, thereby, reducing the solution space.

A knowledge of the existing i*ToNuSMV 2.02 tool is a pre-requirement for understanding the extension provided to the tool in this paper. The URL for accessing this existing version of the tool is as follows: http://cucse.org/faculty/tools/. We will assume that the reader has a sound knowledge of the working of the current tool version (ver2.02) and skip the discussion of some of the basic concepts which have been inherent in all previous versions of the tool. For a brief understanding of the working of the current version of the tool, readers can refer to the user manual given in this link: http://cucse.org/faculty/wp-content/uploads/2017/03/User-Manual_ver2.02.pdf. The rest of the paper is organized as follows: Sect. 2 presents a brief review of the current state-of-the-art for ensuring business compliance in goal modeling. In Sect. 3 we present the basic methodology and assumptions underlying this deployment interface. This is followed by Sect. 4 where we elaborate on the interface of the i*ToNuSMV 3.0 interface with the help of a use case. The section also presents a workflow of the deployment framework. Section 5 presents some experimental data on how the performance of the proposed i*ToNuSMV 3.0 varies with respect to that of the existing version. Finally, we conclude the paper with Sect. 6.

2   Review

Business processes undergo evolutionary change of lifecycle from an unsatisfactory state to desired state. This volatility of business process models presents the need of methods to control and trace the evolution process [11]. There are works that represented business process model in a goal model form and specifications are applied on the goal model to check the validity of the process model. The addition, removal or modification of goals ultimately result in modification of temporal ordering of events. Business processes and their management have always introduced challenges for organizations, as the processes are often cross-functional. In [13], the author has proposed a business process management methodology and framework based on the User Requirements Notation as the modelling language and jUCMNav as the supporting tool for modelling and monitoring processes. Business processes are subject to compliance rules and policies that stem from domain-specific requirements such as standardization or legal regulations. There are instances when the constraint that has to be applied on process data rather than on the activities and events, so checking such constraint should deal with data conditions as well as ordering of events. Knuplesch et al. [10] have proposed a data aware compliant checking methodology as a preprocessing method to enable correct verification. A methodology that directly transforms a business process model into finite state machine and perform compliant checking using linear temporal logic (LTL) has been proposed in [9]. It is mainly focused on verification of control flow aspects rather than verification of the state of data objects. [12] is focused on the fact that deployment efficiency of business process models can be improved by checking the compliance of business process models by model checking technology. The existing literature [12] in this domain highlights the importance of model checking against temporal properties and has come with various ways to do so.

Often business process models are represented using goal models for validating the requirements. Unlike dataflow models, goal models like i* are sequence agnostic which make them prone to errors arising out of non compliance towards temporal properties. Works have been done [2] to bridge this gap by transforming goal models into finite state models. These finite state models can be fed into standard model verifiers like NuSMV and can be checked against temporal properties. The current i*toNuSMV ver2.02 can only check a given goal model against temporal properties [3]. However, if the property is not satisfied it does not identify the state transitions due to which the property fails to get satisfied. Several goal model analysis techniques have been proposed in recent years, some techniques propagate satisfaction values through links to and from goals in the model, others apply metrics over the structure of the model, apply planning techniques using tasks and goals in the model, run simulations over model tasks, and yet others perform checks over model contents [8]. A survey of these different approaches show that model checking using temporal property provides satisfactory result, but an iterative process of manually defining the bounds of the model checker is often required.

In [7], the authors have introduced a goal-oriented requirements engineering method for compliance with multiple regulations. Legal statements are modeled using Legal-GRL and their solutions for a goal model, have been linked with organizational-GRL. The business processes and their operations should satisfy a set of policies or constraints characterized by compliance rules. The change in compliance rules can occur in line with the business goals, and also with legal regulations. [14] has highlighted the importance of legal compliance checking in business process model. Any non-compliance may lead to financial and reputation loss. A review of various goal- oriented frameworks for business compliance checking indicates that more research in this domain is required to face different challenges that comes up with changing compliance rules. The existing framework of model checking only provides a counter example when compliance rules are not satisfied but it does not come up with the exact model that is compliant with the given temporal property.

3 The Proposed Methodology [4]

In this section, the guidelines proposed in [4] for pruning the finite state models generated by the Semantic Implosion Algorithm (SIA) [2] are briefly recapitulated. The assumptions that were listed in the previous work are mentioned first, followed by the different types of CTL properties that are handled.

3.1 Assumptions

The different types of CTL constraints have been studied in detail and this paper works with a finite subset of such constraints in the framework. The primary goal is to generate a compliant finite state model by pruning transitions from the finite state model generated by i*ToNuSMV ver2.02. The proposed guidelines have the following four assumptions:

A-1: Since FSMs are derived for fulfillment of goals, the framework works with only AG and EG temporal operators for the violation of goal fulfillment. Example: AG(V109!=FU).

A-2: Two CTL predicates can be connected through Boolean connectives like AND and OR. This framework allows the user to define only two predicates at a time and connect them by theAND or OR operator. Example: AG(V109!=FU AND V102!=FU).

A-3: Another type of CTL constraint that is addressed is implication (→). Any two constraint can have implication between them. The implication operator has been restricted to only single level of nesting. Example: AG(V101=CNF→AF(V102=FU AND V103!=FU))

A-4: The goal tree level for an actor has been assumed to be 3 to reduce the problem complexity.

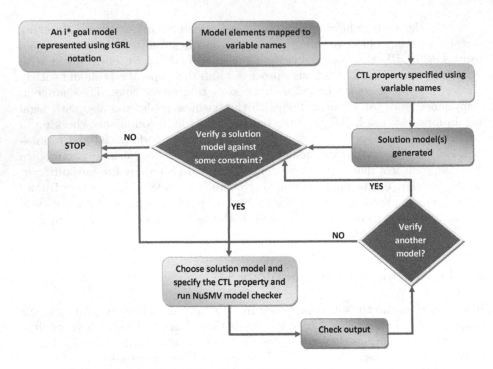

Fig. 1. The workflow of the i^* *ToNuSMV 3.0* deployment framework.

3.2 CTL Properties Handled

This section briefly explains each of the CTL constraints that were addressed in [4] and how the corresponding finite state models are derived.

1. **EG(V#!=FU) for AND decomposition.** This type of properties are safety properties that prevent something bad from happening. Ensuring this property on a goal with AND-decomposition requires the pruning of CNF→FU transitions for some subset of the child nodes.
2. **EG(V#!=FU) for OR decomposition.** The same type of safety property on a goal with OR-decomposition has different consequences. Ensuring such a property requires the pruning of CNF→FU transitions for all the child nodes.
3. **EG(V#!=FU AND V#!=FU).** CTL properties which have multiple CTL predicates connected with boolean AND connectives can be ensured by satisfying each predicate separately. The solution space is a Cartesian product of models generated from each CTL predicate - denoted by $M \times N$.
4. **EG(V#!=FU OR V#!=FU).** CTL properties having multiple CTL predicates connected with boolean OR connectives can be ensured by satisfying either of the predicates or both. The solution space is much larger and denoted by $M + N + (M \times N)$.
5. **AG(V#!=FU→V#!=FU).** These type of CTL properties (defined with the implication operator →) specify an ordering over the fulfilment of goals.

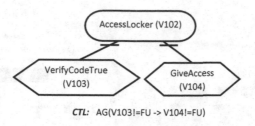

CTL: AG(V103!=FU -> V104!=FU)

Fig. 2. A simple goal model for Accessing a Locker

Thus, all those invalid states need to be pruned from the FSM that violate this property. State transitions to or from these invalid states are correspondingly removed.

6. **EG(V#!=FU) for AND-OR decompositions.** Ensuring such safety properties for multi-level goal models with OR-decompositions nested under an AND decomposition requires the pruning of CNF→FU transitions for all OR-children. This needs to be done for any subset of the AND-children of the root node.

7. **EG(V#!=FU) for OR-AND decomposition.** If the root goal is OR-decomposed followed by each OR-child undergoing an AND-decomposition, then these types of CTL properties can be ensured by pruning CNF→FU transitions for any subset of AND-children for each of the OR-child of the root goal.

The above seven types of CTL properties have been addressed in the newly proposed version 3.0 of the *i* ToNuSMV* framework. For a more detailed understanding of how each of these CTL property classes are ensured within a goal model, readers can refer to [4]. Figure 1 demonstrates the workflow of the *i* ToNuSMV 3.0* framework.

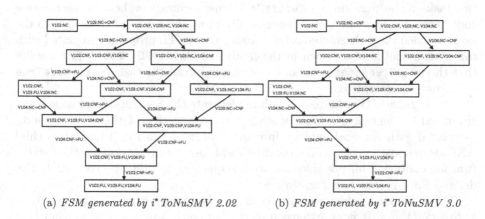

(a) *FSM generated by i* ToNuSMV 2.02*   (b) *FSM generated by i* ToNuSMV 3.0*

Fig. 3. Finite state models derived by the existing and new version of the i*ToNuSMV tool.

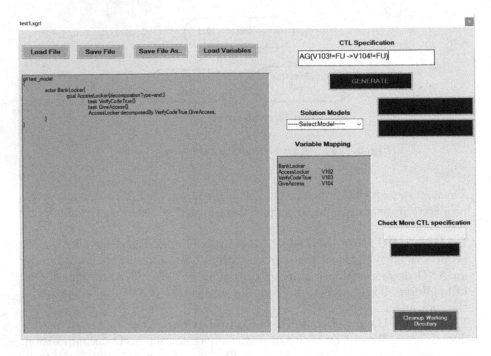

Fig. 4. The input interface of i*ToNuSMV 3.0. Both the goal model and the CTL property are provided as input.

4   Demonstration with Case Study

In this section, the working of the *i** *ToNuSMV 3.0* deployment interface is demonstrated with the help of a simple real life case study. Figure 2 shows a simple goal model that captures the requirements for `Access Locker`. It requires two tasks to be performed - `VerifyCodeTrue` verifies whether the user access code entered is true and `GiveAccess` finally gives the access of the locker to the user provided the code entered is true. An intuitive CTL property associated with this goal model is also shown in the figure. `AG(V103!=FU→V104!=FU)` implies that the task `GiveAccess` cannot be performed until the task `VerifyCodeTrue` is successfully completed.

The *i** *ToNuSMV 2.02* tool, which implements the Semantic Implosion Algorithm (SIA), generates a finite state model irrespective of the CTL property associated with the goal model. Since the goal model in Fig. 2 has a two child AND-decomposition, the corresponding FSM has a 2-dimensional lattice structure for capturing all possible execution sequences to fulfil the root goal. The derived FSM is shown in Fig. 3(a).

The research guidelines proposed in [4] have been implemented in *i** *ToNuSMV 3.0*. It is an extension of the Semantic Implosion Algorithm that takes the finite state model generated by SIA and prunes those transitions which

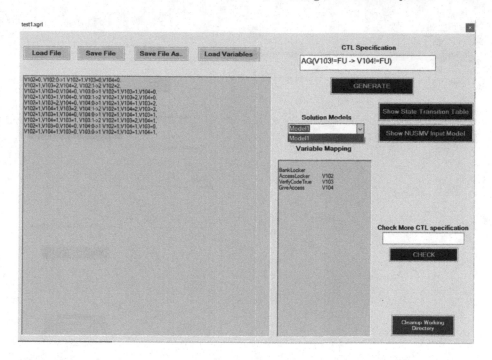

Fig. 5. The output interface of i*ToNuSMV 3.0. All possible finite state models can be selected for verification purposes.

violate the given CTL property. The pruned finite state model for the given goal model and CTL property (refer to Fig. 2) is shown in Fig. 3(b).

4.1 *i** *ToNuSMV 3.0* Input Interface

There are some modifications in the framework interface for *i** *ToNuSMV 3.0* over the existing 2.02 version. Previously, users could load a goal model, generate a finite state model (FSM) and then check the FSM against some CTL property using the integrated NuSMV model checker. In the new version, users can provide a CTL formula along with the goal model as input. The ⟨GENERATE⟩ button generates one or more finite state models that are compliant with the given CTL property. The input interface is shown in Fig. 4.

4.2 *i** *ToNuSMV3.0* Output Interface

Once the user generates all possible finite state models that are compliant with the given CTL property, he/she can view each one of them from within the interface. The ⟨Solution Models⟩ drop-down menu allows the user to choose the particular solution that he/she desires. Like the previous 2.02 version, users can see both the finite state model and the corresponding NuSMV input model of the chosen solution. This output interface is illustrated in Fig. 5.

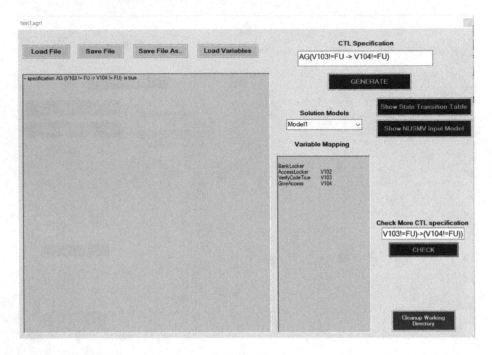

Fig. 6. The verification interface of i*ToNuSMV 3.0. Any output finite state model can be verified or checked against CTL properties.

4.3 *i* *ToNuSMV3.0* **Verification Interface**

Since the *i* *ToNuSMV 3.0* framework claims to generate compliant finite state models, the interface also supports the previously in-built NuSMV model checker for cross-verification purposes. Once the solution models are generated, the user can choose any one of them and re-enter the CTL property (that was provided as input) in the ⟨Check More CTL specification⟩ tab. On clicking the ⟨CHECK⟩ button, the user will always get a TRUE output. This is shown in Fig. 6. The user can also choose to check the pruned finite state model against other CTL properties if desired.

4.4 **URL**

The *i* *ToNuSMV 3.0* framework can be downloaded from the following link: https://github.com/istarToNuSMV/i-ToNuSMV3.0. The User Manual and use case examples have been shown on the webpage.

5 Experimental Results

In this section, some experimental results have been documented that were obtained after performing extensive simulations with the existing (version 2.02)

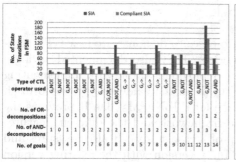

(a) *Number of state transitions in the final FSM*

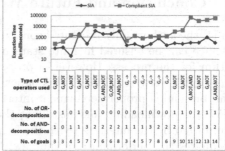

(b) *Execution time for deriving the final FSMs*

Fig. 7. Finite state models derived by the existing and new version of the i*ToNuSMV framework.

and newly proposed (version 3.0) versions of the *i* ToNuSMV* framework. Arbitrary goal models were designed with varying complexity in terms of the number of actors, the number of goals, the number of AND/OR - decompositions and the complexity of the associated CTL constraints. The simulations did not bring out any anomalous behaviour. Data were collected with respect to the number of transitions in the final output FSM and the execution time. These results have been plotted and shown in Fig. 7.

The bar chart of Fig. 7(a) shows a comparative analysis between SIA (implemented in version 2.02) and Complaint-SIA (implemented in version 3.0). *i* ToNuSMV 2.02* does not generate a compliant FSM like *i* ToNuSMV 3.0*. Thus, the FSM generated by version 2.02 includes all possible execution sequences between sets of states. The complaint-FSM generated by version 3.0 will have fewer number of transitions as all CTL properties used in these simulations, impose some sort of ordering between events. This results in the final FSM having only a subset of the transitions included by SIA. The degree (or %) of reduction in state space is dependent on several factors rather than only one.

The line plot shown in Fig. 7(b) compares the execution time of SIA and Complaint-SIA - both measured in milliseconds. With the same set of simulation parameters, it is observed that *i* ToNuSMV 3.0* takes much more time than *i* ToNuSMV 2.02* to generate the finite state models. This is also quite logical as version 3.0 implements some additional checks and tasks after SIA is executed (as in version 2.02). Basically, version 3.0 takes the FSM generated by SIA and individually scans and prunes transitions to satisfy the given CTL property. Also, as discussed in [4], there may be multiple strategies for pruning different subsets of transitions in order to satisfy the CTL property. *i* ToNuSMV 3.0* executes each such strategy and generates a unique finite state model (pruned and compliant) for each of these strategies. This is the reason why version 3.0 takes much longer to reach completion.

6   Conclusion and Future Work

In [4] the authors only presented some use cases to demonstrate how CTL compliance can be ensured in finite state models. They also documented an algorithm for the process. However, a proper deployment framework implementing the guidelines were missing. This paper builds on the guidelines proposed in [4] and presents a proper deployment interface for i*ToNuSMV 3.0. It provides the URL for downloading and installing the framework and the different features supported by the interface (with the help of a case study). It also measures and compares the performance of the newly built version with the existing version of the i*ToNuSMV tool (also see Table 1).

Table 1. Feature comparison between Verions 2.02 and 3.0

Features	i*ToNuSMV 2.02	i*ToNuSMV 3.0
Input specification	i* goal model defined using tGRL	i* goal model defined using tGRL and a CTL property
Number of FSM generated	1	1 or more than one
Compliant FSM	FSM may or may not be compliant to any temporal property	FSM compliant with a given CTL property
Solution space	Comparatively large	Reduced solution space
Number of NuSMV input	1	One for each of the FSM generated
Verification	NuSMV model checker verifies property on single finite state model	NuSMV model checker can separately verify each of the solution model

This work presents a tool to generate finite state models from goal models that already satisfy some given CTL constraint. Finite state models can be more readily transformed into code. Thus, this research takes an important step towards the development of business compliant applications directly from goal models. Most business compliance rules have some sort of temporal ordering over events and can be represented with temporal logics efficiently. However, the proposed solution has several assumptions which needs to be relaxed for making the framework more complete.

One of the more important limitations of the proposed solution is that only one temporal property (in CTL) can be specified along with the goal model specification. Future versions of the *i*ToNuSMV* framework will aim to allow users to specify multiple CTL properties over a single goal model specification. Another limitation of the new version is the extra processing time that is required. The additional pruning mechanism requires extensive checking of the finite state model generated by SIA. Currently research efforts are being channelized to develop an efficient version of the Semantic Implosion Algorithm that will generate compliant FSMs in a more efficient manner.

Acknowledgement. This work is an extension of the Ph.D. work of Novarun Deb, who was a research fellow in the University of Calcutta under the Tata Consultancy Services (TCS) Research Scholar Program (RSP). We acknowledge the contribution of TCS Innovation Labs in funding this research. This work has also been partially supported by the Project IN17MO07 "Formal Specification for Secured Software System", under the Indo-Italian Executive Programme of Scientific and Technological Cooperation.

References

1. Deb, N., Chaki, N., Ghose, A.K.: Using i* model towards ontology integration and completeness checking in enterprise systems requirement hierarchy. In: IEEE International Model-Driven Requirements Engineering Workshop (MoDRE) (2015)
2. Deb, N., Chaki, N., Ghose, A.K.: Extracting finite state models from i* models. J. Syst. Softw. **121**, 265–280 (2016). https://doi.org/10.1016/j.jss.2016.03.038
3. Deb, N., Chaki, N., Ghose, A.K.: i*ToNuSMV: a prototype for enabling model checking of i* models. In: IEEE 24th International Requirements Engineering Conference (RE) (2016)
4. Deb, N., Chaki, N., Roy, M., Bhaumik, A., Pal, S.: Extracting business compliant finite state models from i* models. In: 6th International Doctoral Symposium on Applied Computation and Security Systems (ACSS) (2019, accepted)
5. Fuxman, A., Liu, L., Pistore, M., Roveri, M., Mylopoulos, J.: Specifying and analyzing early requirements: some experimental results. In: Proceedings of the 11th IEEE International Requirements Engineering Conference, pp. 105–114 (2003). https://doi.org/10.1109/ICRE.2003.1232742
6. Fuxman, A., Pistore, M., Mylopoulos, J., Traverso, P.: Model checking early requirements specifications in Tropos. In: Proceedings Fifth IEEE International Symposium on Requirements Engineering, pp. 174–181 (2001). https://doi.org/10.1109/ISRE.2001.948557
7. Ghanavati, S., Rifaut, A., Dubois, E., Amyot, D.: Goal-oriented compliance with multiple regulations. In: IEEE 22nd International Requirements Engineering Conference (RE) (2014). https://doi.org/10.1109/RE.2014.6912249
8. Horkoff, J., Yu, E.K.H.: Analyzing goal models: different approaches and how to choose among them. In: SAC (2011)
9. Kherbouche, O.M., Ahmad, A., Basson, H.: Formal approach for compliance rules checking in business process models. In: IEEE 9th International Conference on Emerging Technologies (ICET) (2013). https://doi.org/10.1109/ICET.2013.6743500
10. Knuplesch, D., Ly, L.T., Rinderle-Ma, S., Pfeifer, H., Dadam, P.: On enabling data-aware compliance checking of business process models. In: Parsons, J., Saeki, M., Shoval, P., Woo, C., Wand, Y. (eds.) ER 2010. LNCS, vol. 6412, pp. 332–346. Springer, Heidelberg (2010). https://doi.org/10.1007/978-3-642-16373-9_24
11. Koliadis, G., Ghose, A.: Relating business process models to goal-oriented requirements models in KAOS. In: Hoffmann, A., Kang, B., Richards, D., Tsumoto, S. (eds.) PKAW 2006. LNCS (LNAI), vol. 4303, pp. 25–39. Springer, Heidelberg (2006). https://doi.org/10.1007/11961239_3

12. Negishi, Y., Hayashi, S., Saeki, M.: Establishing regulatory compliance in goal-oriented requirements analysis. In: IEEE 19th Conference on Business Informatics (CBI) (2017). https://doi.org/10.1109/CBI.2017.49

13. Pourshahid, A., et al.: Business process management with the user requirements notation. Electron. Commer. Res. **9**, 269–316 (2009)

14. Ghanavati, S., Amyot, D., Peyton, L.: A systematic review of goal-oriented requirements management frameworks for business process compliance. In: IEEE Fourth International Workshop on Requirements Engineering and Law (2011). https://doi.org/10.1109/RELAW.2011.6050270

15. Yu, E.: Modelling strategic relationships for process reengineering. Ph.D. thesis, University of Toronto, Toronto, Canada (1995)

A Methodology for Root-Causing In-field Attacks on Microfluidic Executions

Pushpita Roy[1,2]([✉]), Ansuman Banerjee[1], and Bhargab B. Bhattacharya[3]

[1] Indian Statistical Institute, Kolkata, India
{pushpita,ansuman}@isical.ac.in
[2] Calcutta University, Kolkata, India
[3] Indian Institute of Technology, Kharagpur, India
bhargab.bhatta@gmail.com

Abstract. Recent research on security and trustworthiness of micro-fluidic biochips has exposed several backdoors in their established design flows that can lead to compromises in assay results. This is a serious concern, considering the fact that these biochips are now extensively used for clinical diagnostics in healthcare. In this paper, we propose a novel scheme for root-causing assay manipulation attacks for actuations on digital microfluidic biochips that manifest as errors after execution. In particular, we show how the presence of a functionally correct reaction sequence graph has a significant advantage in the micro-fluidic context for debugging errors resulting out of such attacks. Such a sequence graph is the basis from which the actuation sequence to be implemented on a target Lab-on-chip is synthesized. In this paper, we investigate the possibility of using this sequence graph as a reference model for debugging erroneous reaction executions with respect to the desired output concentration. Our debugging method consists of program slicing with respect to the observable error in the golden implementation. During slicing, we also perform a step-by-step comparison between the slices of the erroneous output with other erroneous and error-free outputs. The reaction steps are then compared to accurately locate the root cause of a given error. In this paper, we consider two different types of assay descriptions, namely (a) unconditional assays, which have a fixed execution path, and (b) conditional assays that alter the execution at runtime depending on the outputs of sensor observations. Experimental results on the Polymerase Chain Reaction (PCR) and Linear Dilution Tree (LDT) and its conditional variant show that our method is able to pinpoint the errors.

1 Introduction

Microfluidic lab-on-chips (LoC) are set to replace cumbersome laboratory-based procedures, and are being considered as the next generation platform for on-chip implementation of biochemical laboratory assays [26]. The benefits of these devices have already been established in myriads of application domains, in DNA analysis, toxicity grading, molecular biology, drug design, rapid and accurate diagnosis of various diseases including malaria, human immunodeficiency

© Springer-Verlag GmbH Germany, part of Springer Nature 2020
M. L. Gavrilova et al. (Eds.): Trans. on Comput. Sci. XXXV, LNCS 11960, pp. 119–152, 2020.
https://doi.org/10.1007/978-3-662-61092-3_7

virus, and for mitigating neglected tropical diseases prevalent in the developing countries [21].

Digital micro-fluidic (DMF) design automation tools take in a biochemical assay description expressed as a reaction sequence graph and translate it into a sequence of actuation operations for a target LoC architecture, which is loaded onto an on-chip controller, for further generation of actuation sequences for enabling various fluidic operations. During reaction execution on the LoC, embedded sensors placed below the electrodes [16] along with overhead CCD cameras [19] are used to monitor reaction progress. Several automation methodologies and frameworks have been proposed to enable complete design flows from assay descriptions to LoC control [3,9,11–13,17,18,27–32].

The evolving landscape of DMF biochips (DMFB) is confronted with a new threat today due to concerns of security attacks and malicious manipulations. While there has been substantial research on the different phases of DMFB design, synthesis and optimization, security threats arising in this context have been relatively quite less studied. A recent work [1] in this direction has analyzed and identified several alarming backdoors in the different phases of the DMFB design life cycle that can be compromised by an attacker, thereby leading to undesirable consequences. Attacks ranging from assay manipulations, protocol modifications, actuation sequence tampering, routing path alterations, sensor data manipulations, mixer timing modifications, artificial transport delay injections, and other device level threats have been identified as serious security issues and countermeasures are being worked on. [2] presents a comprehensive survey of different attack possibilities that can lead to breaches in the DMFB context. [23] presents a categorization of assays to find a generic model for error detection to combat attacks.

The objective of this paper is to propose a method to debug errors resulting out of attacks in the DMFB design flow, and propose methods for automated identification and attack site localization. In particular, we consider that our method can be applied in situations where an attacker modifies the mixer mixing cycle times, or artificially injects path delays in the droplet routes, with an objective of jeopardizing the expected execution timing of a given assay when executed on a given DMF biochip. These attacks can be administered in a variety of ways, by modifying the actuation sequence, by manipulating the mixer cycle times, inserting additional mix/move operations, lengthening droplet movements or by making device level modifications. We assume that the final manifestation of the timing attack is observable as errors through some sensor observation at some point, either at the outputs or at intermediate observation points through sensors fitted underneath each electrode or through overhead CCD cameras. These timing modifications may have different consequences. As an example, a delay in droplet movement can result in a delay in its arrival at a mixer input electrode, failing which the mixer may be instantiated incorrectly without the required number of droplets on its input pins, thereby resulting in volumetric errors. Similarly, a mixer instantiated earlier than specified will lead to incorrect mixing. Another example may be of mixer cycle time modifications, for which the mixer may work for less/more time, thereby jeopardizing the operations which depend on its output droplets. The severity of the consequences of these attacks has inspired new research directions in recent times that can detect, counter and isolate these threats and ensure smooth DMFB operations.

Faults or attacks of the above kind are runtime manifestations. The actuation sequences synthesized from the golden implementation of the biochemical description for a target architecture are usually not fault-tolerant, and may thus lead to errors after completion of assay execution on the faulty LoC. The errors manifest as incorrect output volumetric concentration at the end of execution or at intermediate detector locations. This needs to be debugged and fixed or isolated with respect to the desired output concentration factor specified as part of the golden reaction, before the next reaction is executed on the erroneous LoC.

In this paper, we study the problem of root-causing execution errors that result from assay attacks in protocol executions. In particular, we propose here an algorithmic method for error identification and error origin localization. Our approach takes in a golden implementation of a reaction sequence graph and error manifestations, and attempts to root cause the faulty operation and location on the LoC architecture. As output, our method produces a set of suspected locations and operations that may be possible origins of the fault. We address the debugging problem for two different types of assay descriptions, namely (a) unconditional assay protocol graphs that have a fixed execution path, and (b) conditional assay graphs, that have different execution paths depending on the outcome of certain observations at intermediate checkpoints. These assays are summarized with an if-then-else structure, that effectively expresses the conditional execution. This support is not available in unconditional assay protocol graphs, that have been the major focus of study in microfluidic research. We present a description of the execution semantics of such conditional protocol graphs, and address the debugging problem for these as well.

The foundation of our debugging approach is based on the concept of program slicing [4], a popular technique used in software engineering parlance for debugging program errors. Our proposed method includes two main steps. In the first step, we slice the assay description with respect to the error being debugged, to eliminate all those operations from the assay which do not affect the erroneous output directly or indirectly. This cuts down the suspected error region significantly since the assay may include many other operations which are not relevant to the fault manifestation, and thereby, do not help in the root causing exercise. In the second stage, we make use of other erroneous manifestations and error free outputs, if any. We simultaneously compute the slices of these, and compare them with the slice of the erroneous manifestation being debugged, to further localize the error source. A naive debugging method would involve debugging the entire assay execution on the LoC architecture, which may be an arduous time-consuming task for a reasonably complex protocol for a moderately sized LoC. Our method significantly cuts down the region of the assay to be examined.

Error correcting methods for cyberphysical integration have been proposed in [19]. These methods aim to identify suspected erroneous regions online during an error occurrence in execution, and propose re-synthesis of the suspected error regions to include operations that can affect the detected fault. We believe that our method can complement such approaches and help in localizing the fault even better, and lead to lesser number of assay operations for re-synthesis. Defects in flow-based micro-fluidic chips have been studied in [14,15], where a fault diagnosis technique using syndrome analysis is proposed. The objective of our paper is to explore the attack root-causing problem in the DMF context. The problem of fluidic constraint violation checking that is addressed in [5] is not useful in our problem setting, since these methods work side-by-side with synthesis tools, and do not consider execution errors in the LoC. The work in [22] discusses the problem of debugging security attacks using dynamic slicing. Authors in [33] use the notion of program slicing for inserting detection operations on a given input sequence graph and re-executing the subprogram which may cause the erroneous output. In contrast, our method not only identifies the set of operations of the given assay that may cause erroneous outputs but also isolates the particular operations that are erroneous. In [24], we have presented an early model for attack debugging and error localization for unconditional assays. This paper is a much enriched variant that handles more general assay descriptions.

To demonstrate the working of our method, we experimented on models of the Polymerase Chain Reaction (PCR) [25] and the Linear Dilution Tree (LDT) [6] protocol along with its conditional variant, with synthetically created random attacks. In both the cases, we were able to localize the attack source, as elaborated in Sect. 5.

This paper is organized as follows. Section 2 presents the problem statement. Section 3 presents the methodology for unconditional assay protocols, while Sect. 4 presents the same on conditional assays. Section 5 presents the details of our implementation and experiments, while Sect. 6 concludes the discussion.

2   Problem Statement

Formally, we are given the following.

- An input sequence graph of a bio-chemical protocol
- The actuation sequence synthesized for a target DMF
- A set of erroneous outputs (at least one), and
- A set of error-free outputs (if present).

Our objective is to find the possible root cause locations of the error in the assay. For the sake of simplicity, we consider a single attack point in this work. In other words, the origin of the attack is a single location or operation or resource, while the manifestations may be errors at multiple outputs.

In this paper, we consider two different types of bio-chemical assays, as below.

- Unconditional Assays: These have a fixed execution sequence as outlined in the actuation sequence synthesized from the assay. For such assays, the instructions are specified as a sequence of instructions, some of which can be executed concurrently, as long as dependencies are preserved. All instructions as specified in the assay and the synthesized actuation sequence are executed by default, without any decisions at run-time. These assays have been the usual artifact of study in most of microfluidics research.
- Conditional Assays: These are unconditional assays augmented with a conditioned execution facility, that can be used to specify conditional execution for certain operations. More specifically, a set of assay statements in such assay descriptions can be guarded by a *condition check* instruction, evaluation of which at run-time determines the exact path the assay is going to take. This is similar to the if-else structure in high level programming languages. If the condition node evaluates to true at run-time based on sensor observations, the set of operations enclosed in the if block are taken up for execution, while the set of operations in the else block are used in case the conditional evaluates to false. Evidently, synthesis of such conditional assay graphs is much different from the standard synthesis procedures for unconditional graphs as reported in literature. We present a detailed discussion on such assay graphs in Sect. 4.

As we discuss in the following sections, the target objective is the same for both the types of graphs, in particular, we attempt to root-cause sources of attacks. The basis of our diagnosis procedure is similar as well, however, the methodology for conditional assays is a little more involved, due to the extra flexibility that they bring in. In the following, we illustrate the solution methods for each.

3   Debugging Errors in Executions of Unconditional Assays

In this section, we first describe the methodology for debugging unconditional assays. As mentioned earlier, all operations described in these assays are executed by default, without any decision or alteration at run-time. We provide below an example of one such assay along with a description of the attack manifestation.

Example 1. Consider the input bioassay P as shown in Fig. 1. Two reagents (as depicted by nodes R_1 and R_2), are received as inputs to a mix operation M_1, where they are mixed for 4 units of time, starting from time unit 5, ending at 9. It may be assumed that time units 1–4 are spent in dispensing and moving the dispensed droplets to the input ports of the mixer. One of the outputs produced after the mix-split operation M_1 is then sent as input to the mix operation M_4, along with a newly dispensed droplet, shown in the Figure as R_3. The mix operation M_4 continues from time unit 15 to 19, thereby producing 2 output droplets at the end of the mix-split operation. The rest of the operations are similar in nature, and have similar interpretations. R_is represent the dispensing of reagents,

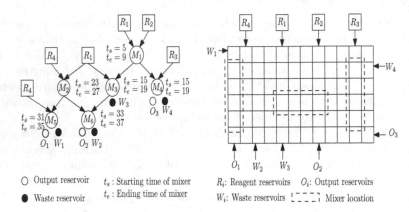

Fig. 1. Input sequence graph of an input bioassay and a target grid. (Color figure online)

M_is the mix operations, and O_is the outputs. In this bio-assay, the execution of the operations do not depend on any run-time decisions. Consider a biochip of dimension 6×12 on which the protocol is executed. Reagent reservoirs are attached with the cells $(1,2)$, $(1,5)$, $(1,8)$ and $(1,11)$, the waste reservoirs are with the cells $(3,1)$, $(6,3)$, $(6,5)$, and $(3,12)$ and the output reservoirs are with the cells $(6,1)$, $(6,8)$ and $(6,12)$.

Let us assume after execution of the above assay, an erroneous output (incorrect concentration/volumetric ratio) is detected at the output reservoir $(6,8)$. The erroneous outputs O_1 and O_2 are shown in Fig. 1 as red circles, while the other output O_3 is error-free. We attempt to find the root cause of the error manifestation in O_2 at $(6,8)$. For this, we need to debug the erroneous output to identify the source of the error. All operations that either directly or transitively affect the erroneous output are suspected error sources. If there are other error free outputs, the operations which directly or transitively affect them can be excluded from the set of suspected error sources, since we assume a single attack point model. Further, operations affecting multiple erroneous outputs can be analyzed to isolate the common operations as the set of most probable root causes. ■

In the microfluidic life-cycle, assay descriptions are executed on a target grid, for which the actuation sequences are synthesized. We explain this on the example above. Let us assume that the example in Fig. 1 is realized on a DMFB grid of size 6×12. Figure 2 shows the actuation sequence of the assay given in Fig. 1. According to Fig. 1, droplets of reagents R_1 and R_2 are mixed by mix operation M_1. In the actuation sequence, droplets of reagents R_1 and R_2 are dispensed at time instant $t = 1$ at locations $(1,5)$ and $(1,8)$ respectively, as shown in Line 1 (d stands for dispense operation). Then in Line 2, the droplet of R_1 moves from location $(1,5)$ to $(2,5)$ (m stands for move operation) and the droplet of R_2 moves from location $(1,8)$ to $(2,8)$ at time instant $t = 2$, and so on. In Line no 5, two droplets at locations $(4,5)$ and $(4,8)$ are mixed for 4 time cycles by a 1×4 mixer (mixsplit is used to depict mix operation followed by a split operation). After the mixsplit operation, two output droplets are generated at locations $(4, 5)$ and $(4, 8)$ at time instant $t = 9$. These two droplets again start to move to their intended target locations at time instant $t = 10$. In this example, all the operations are performed as specified, no runtime decision is needed to perform any operation.

```
1.  d(R₁,1,5) d(R₂,1,1,8)
2.  m[(1,5) -> (2,5)] m[(1,8) -> (2,8)]
3.  m[(2,5) -> (3,5)] m[(2,8) -> (3,8)]
4.  m[(3,5) -> (4,5)] m[(3,8) -> (4,8)]
5.  mixsplit(4,5,4,8,4)
10. m[(4,5) -> (4,6)] m[(4,8) -> (4,9)]
11. m[(4,6) -> (4,7)] m[(4,9) -> (4,10)]
12. d(R₁,1,1,5) d(R₃,11,1,11)
· · ·
· · ·
```

Fig. 2. Actuation sequence of the assay with straight execution

3.1 Debug Methodology

Our debugging method consists of 3 main steps, as described below.

- *Time frame expansion*: This creates a time annotated reaction graph based on the given reaction graph and the actuation sequence on the target LoC.
- *Slice computation*: This helps in pruning a given reaction graph to discard operations irrelevant to a given operation.
- *Error localization*: This involves comparison between the slices for more accurate localization.

Time Frame Expansion: The objective of this step is to instantiate the given reaction graph (e.g. in Fig. 1) on the biochip architecture, and unfold the actuation sequence over time, thereby creating a graph with time annotated operations, as in Fig. 1. Figure 3 shows the transformed input sequence graph, where the square boxes represent the set of dispense operations and circular nodes represent the set of operational vertices of the assay. An edge from one vertex to another represents a dependency between them. Some nodes are annotated with a time interval which depicts the start and end time of the operation. Each reagent instance in Fig. 1 is replaced with a corresponding dispense operation with a unique identifier $(D_1, D_2, \ldots, D_4)$. Each mix operation is also uniquely identified by an identifier $(M_1, M_2, \ldots, M_6)$. The figure shows another type of circular node labeled as mv, which represents a move operation that depicts the movement of a droplet from a location (x, y) to another location (x', y') on the

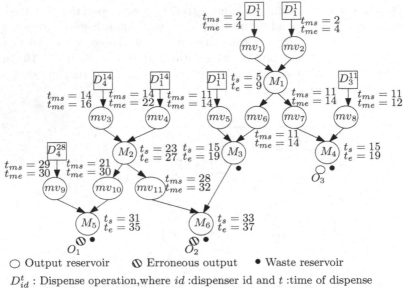

○ Output reservoir   ◐ Erroneous output   • Waste reservoir

D_{id}^{t} : Dispense operation,where id :dispenser id and t :time of dispense
mv :move operation from source to destination M: mix operation
t_{ms} : Start time of move operation; t_{me} :End time of move operation
t_s : Start time of mix operation; t_e :End time of mix operation

Fig. 3. Unfolded Input sequence graph of an input bioassay.

DMF grid. Each operation in the input sequence graph is annotated with the start and end times. The time frame expansion of the original reaction graph gives us a better artifact to debug, and we use it in our method. The output erroneous droplets are shown in red in Fig. 3.

Slice Computation: The objective of the slicing operation is to prune out all operations that have no influence on the erroneous output. The notion of a slice [4] has been extensively used in the software engineering community. We adopt the concept of the slice in the DMF context and show how it serves as an effective debugging aid. Given a time frame expanded graph and an error location, we compute the backward slice to collect all nodes (i.e. operations in the graph) that have a direct or transitive influence on a given node. The backward slice computation attempts to compute the transitive closure of the dependencies encountered in the path from the slicing point to the start of the assay. The first step in slice computation is to identify a slicing criterion. We first consider the error manifestation output as the slicing criteria and compute its backward slice as below.

The inputs to the slicing step are:

– A time frame expanded input sequence graph of an assay P
– A Slicing criteria, in this case, an erroneous output.

Algorithm 1. sliceComputation(*c*)

Unmark all operations in the assay and initialize slice to ϕ

for *all operations OP in the given assay executed before slicing point c* **do**

    if (*OP = dispense*) **then**

        destLocation = dispense(*time*)

        if (*destLocation = c*) **then**

            Add operation *OP* in the stack *S* ;

            Return;

        end

    end

    if (*OP = move*) **then**

        destLocation = move(*src, time*)

        if (*destLocation = c*) **then**

            Add operation *OP* in the stack *S*

            Call sliceComputation(*src*)

        end

    end

    if (*OP = mix*) **then**

        destLocation = mix(*input1, input2, time*)

        if (*destLocation = c*) **then**

            Add operation *OP* in the stack *S*

            Call sliceComputation(*input1*)

            Call sliceComputation(*input2*)

        end

    end

end

Print all marked statements in the assay as the slice

This step produces as output a fragment of the assay *P* that is likely to be the source for the observed error at the output reservoir. Algorithm 1 presents the algorithm for computing the slice for a given assay *P*. We explain its working philosophy. The algorithm sliceComputation takes one argument *c* as input, where, *c* is the slicing criteria, initialized as the erroneous output. The algorithm traverses the time frame expanded graph and checks all operations of the assay that could be executed before the slicing point *c* and pushes those operations on a stack *S* that stores all operations that can affect the slicing criteria directly or indirectly. The operations on the assay are dispense, move and mix. The *dispense* (*time*) operation dispenses a droplet on a cell at a certain time and produces a droplet at the dispense location as output. The *move*(*src, time*) operation takes two arguments: *src* refers to the source location and *time* refers to the time required for the operation and produces a droplet at the *destLocation* as output. The *mix*(*input1, input2, time*) operation takes three arguments, where *input1*, *input2* refer to the location of the two input droplets and *time* refers to the time instance of the operation. It produces 2 output droplets as a result of the mix-split operation. For each of these droplets, the mix operation constitutes the slice. For a mix operation, the two input port locations constitute the slice. For each such input port, the corresponding move operation (which produces a droplet at that location) or a dispense (which dispenses a droplet at that port location) is part of the slice. Similarly, for a move operation, considering the *destLocation* as the slicing criterion, the source constitutes the slice. If *destLocation* matches with the slicing criteria c, then the operation is added to the slice and the algorithm sliceComputation is recursively called for the input location of that added operation. The slice computation is invoked starting from the slicing criterion c. The objective is to mark all nodes in the assay which affect c directly or transitively. To this effect, c is marked, and the slice of c is computed first. Nodes in the slice of c are marked and pushed onto a stack, since these need to examined for slice computation now. This is done by recursively invoking the slice computation on them, as earlier, by popping them off the stack in order and then marking and pushing their slices. The recursion terminates when a dispense method is encountered. In the process, Algorithm 1 is able to collect all operations in the assay that affect the slicing criterion. It is easy to see that nodes in the graph which do not have a direct or transitive influence on c will *not be marked* in any step and thus we have a much smaller fragment of the assay to look at, for possible error localization. The slices for outputs O_1, O_2 and O_3 are shown in Fig. 4(A), (B) and (C).

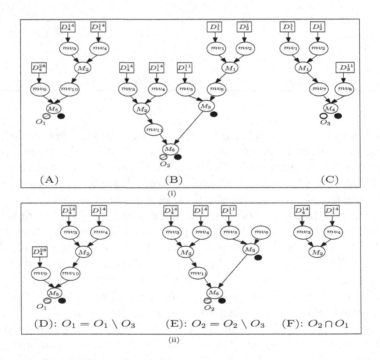

Fig. 4. (i) Slice computation, (ii) Error Localization steps

An illustration of the resulting slice follows below. We explain the slice building process and assay pruning operation with respect to O_2. Figure 5 shows the slice for O_2. The top level node showing a *volumetric error at output O_2* is the root of the slice. The volumetric error at O_2 occurs either due to *the incorrect detection at an output reservoir* or due to *the unsuccessful completion of the mixer M_6*. The event *unsuccessful completion of mixer M_6* occurs either due to an *Error at M_6* or due to the *Absence of the input droplets at M_6*. The event *Absence of the input droplets at M_6* occurs either for the event *Absence of the input droplet 1 at M_6* or for the event *Absence of the input droplet 2 at M_6*. The event *Absence of the input droplet 1 at M_6* occurs either for the event *Error at move operation* or for the event *Unsuccessful completion of mixer M_2*. Another event *Absence of the input droplet 2 at M_6* occurs either for the event *Unsuccessful completion of mixer M_3* or for the event *Error at the move operation*. Continuing similarly, we finally get the slice for O_2 in Fig. 5.

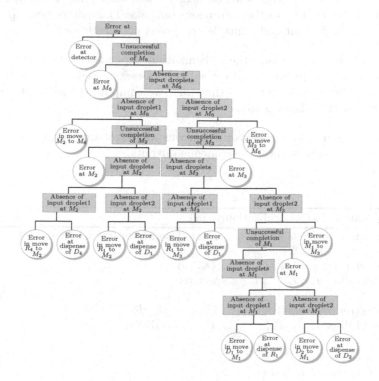

Fig. 5. Slicing in action for o_2

Error Localization: Once the slice for the erroneous output is computed, we have a subset of assay operations affecting the error, and any operation in this subset can be a potential source of the error. Once the slice computation is over, the designer can be provided the marked statements in the resulting slice to examine the root cause of the error. Evidently, there may still be quite a few statements in the slice that the designer needs to look at. We now propose a couple of further optimization steps that can cut down the region even lesser. For this, we assume a *single error source model*, which has been a popular fault model in the electronic design automation industry. We show that it leads to more efficient localization. We build the slices for all assay outputs. This may include other erroneous outputs and error-free outputs. The main motivation is as follows.

- An operation common between an erroneous output and an error-free output cannot be a potential source, since it would have otherwise affected the error-free output as well. Thus, all such operations can be safely *discarded* from the slice of the erroneous output to have a smaller error suspect region.
- Since we have a single source of error, operations common between the slices of erroneous outputs are potential candidates as the source. We can further prune the slice of the error being debugged by *including* only those operations which are common with other erroneous outputs. This cannot be empty since we have a single fault and multiple erroneous manifestations.

For both the cases above, the optimization depends on the availability of appropriate outputs (erroneous or error-free). In case we have multiple manifestations being debugged right from the start, we compute the slices of each and then carry out the following steps:

- Difference computation between erroneous and error-free outputs.
- Intersection between slices of erroneous outputs.

Algorithm 2. ErrorLocalization(n, m)

Erroneous set consists of $\{O_1, O_2, \ldots, O_m\}$ and error free set consists of $\{O'_{m+1}, O'_{m+2}, \ldots, O'_n\}$;

for *all erroneous outputs* $O_1, O_2 \ldots O_m$ **do**
  | build slices $T_{e_1}, T_{e_2}, \ldots T_{e_m}$ respectively;
end
for *all error free outputs* $O'_{(m+1)}, O'_{(m+2)} \ldots O'_n$ **do**
  | build slices $T'_{e_{(m+1)}}, T'_{e_{(m+2)}}, \ldots T'_{e_n}$ respectively;
end
for $i = 1$ *to* m **do**
  | **for** $j = m + 1$ *to* n **do**
  | | $T_{e_i} = T_{e_i} \setminus T'_{e_j}$;
  | **end**
end
for $i = 1$ *to* m **do**
  | $E = \bigcap T_{e_i}$;
end
Print E as the probable error locations;

Algorithm 2 describes the procedure for error localization, that accepts two parameters, the number of outputs (m) and the number of erroneous outputs (n) of the assay. The set of outputs of the assay is divided into two groups, the set of erroneous outputs $\{O_1, O_2, \ldots, O_m\}$ and the set of error free outputs $\{O'_{m+1}, O'_{m+2}, \ldots, O'_n\}$. We build the slices for each erroneous output and for each error free output. Generally, the slice for each erroneous output contains all possible candidate error sources. We then update the slice of each erroneous output by discarding the common operations that are present in the slices of both the erroneous outputs and the error-free outputs, as shown in steps 8 to 10 of Algorithm 2. Again, in steps 11 to 13, we find the intersection of all the updated erroneous output sets for a further optimization. We finally print the resulting slice E as the probable error sources. Figure 4(A), (B) and (C) show the slices for outputs O_1, O_2 and O_3 respectively, in which, O_1 and O_2 are the erroneous outputs and O_3 is the error-free one. We want to debug the erroneous output O_2. Figure 4(D) shows the resultant slice of O_2 after the difference computation with the error-free output O_3. There are two erroneous outputs in the given assay. We compute the final slice as the non-empty intersection of the resultant slice of output O_2 with the other erroneous output O_1 to produce the probable error location. Figure 4(E) shows the probable error location and the output of our method ErrorLocalization and the final slice with 5 nodes.

4   Debugging Conditional Assays

Conditional assays consist of a set of concurrent operations executed in sequence along with some checkpoint observations. Based on the checkpoint observations, some online decisions are taken. For such assays, two sets of instructions are loaded into the biochip controller for execution with each checkpoint observation. Each checkpoint observation is checked with a certain condition. If the condition is satisfied, then a certain set of instructions are executed, else another set of instructions are executed. There are some branching operations in case of conditional assays. To the best of our knowledge, there does not exist much formalization of such conditional assays in literature. In this paper, we first discuss the execution semantics of such conditional assays. We illustrate the notion of conditional assays with the following example.

Example 2. Consider the bio-chemical protocol description shown in Fig. 6. In this bio-chemical protocol, some run-time decisions are taken based on checkpoint observations. In the given protocol, two reagents R_1 and R_2 are mixed using a mix operation M_1 and two output droplets are produced. One output droplet moves to the output reservoir and is observed as a final output FO_1. The other droplet generated out of this mix operation is subjected to a check. Some parameter of this droplet O_1, like concentration of O_1, is measured and checked to see if it is less than some pre-defined δ. The condition checking on an output is shown by an elliptical ring in Fig. 6. If the condition is not satisfied, the output O_1 is mixed with the reagent R_1 by the mix operation M_6. If the condition is satisfied, then the

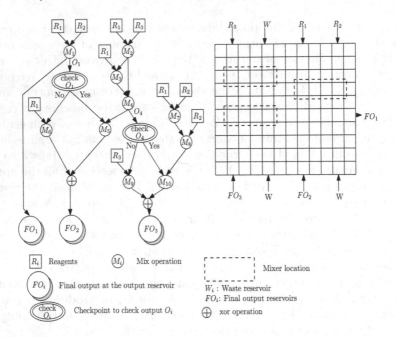

Fig. 6. Input sequence graph of a conditional assay with run-time decisions (Color figure online)

output O_1 is mixed with the output of the mix operation M_4 by the mix operation M_5. Hence, based on the observation of the output O_1 at the checkpoint, either the output of the mix operation M_5 or M_6 is the final output (as depicted with a $\oplus$ node) and is observed at the output reservoir FO_2. On the other hand, the output of the mix M_4 is obtained after performing a set of mix operations, like, the reagents R_1 and R_3 are mixed by the mix operation M_2, then the output of M_2 is again mixed with the reagent R_1 by the mix operation M_3 and the outputs of the mix operations M_2 and M_3 are mixed by the mix operation M_4. Now, the output O_4 of the mix operation M_4 is also checked to examine if the concentration of O_4 is less than some predefined ρ. If the condition is not satisfied, the output O_4 is mixed with the reagent R_3 by the mix operation M_9. If the condition is not satisfied, the output O_4 is mixed with the output of the mix operation M_8 by the mix operation M_{10}. Hence, one among the output of the mix operation M_9 or the output of the mix operation M_{10} is the final output (as depicted with a $\oplus$ node) and observed at the output reservoir FO_3. The output of the mix operation M_8 is produced after performing the mix operation M_7. Here, M_7 mixes the reagents R_1 and R_3 and the mix operation M_8 mixes the output of M_7 with the reagent R_2.

After execution of the bio-assay, the outputs are stored at their corresponding output reservoirs. Let us assume that the output at the output reservoir FO_3 is erroneous. In Fig. 6, the erroneous output is shown as a red colored shaded circle. We need to identify the source of the erroneous output. Firstly, we may *observe that all the operations which directly or indirectly affect the erroneous output FO_3 are the suspected error sources. However, there are two more outputs FO_1 and FO_2, produced at the end of execution which are error-free. Some operations present in the set of operations which are suspected as the source of error of the output FO_3, are also present in the set of operations that lead to the computation of the error-free outputs. Hence, as earlier, those operations can be excluded from the suspected error source. In a later discussion, we elaborate on our debug methodology for identifying the root cause of error manifestations in such conditional assays.* ∎

As mentioned earlier, not much research has been reported in literature on work on such conditional assays. The usual methods of synthesis, routing and waste management that are well understood for unconditional assays have not been much looked at, in the conditional context. More importantly, the structure of the actuation sequence synthesized for such conditional assay descriptions are quite different, and outside the purview of the available synthesis tools to generate, as discussed in the following.

Example 3. Consider that the example shown in Fig. 6 is realized on a DMFB grid of size 9×12. The reagent reservoirs R_1, R_2 and R_3 are placed at locations $(1,8)$, $(1,11)$ and $(1,2)$ respectively. The output reservoirs FO_1, FO_2 and FO_3 are placed at locations $(6,12)$, $(9,8)$ and $(9,2)$ respectively. There are three 1×4 mixers placed on locations $(3,2,3,5)$, $(4,8,4,11)$ and $(6,2,6,5)$ respectively. After synthesis of the given conditional assay on the 9×12 DMFB grid, we derive an actuation sequence, as shown in Fig. 7. Figure 7 consists of a few rectangular boxes, where each such box consists of a set of instructions. A set of concurrently executing instructions are depicted on each line, prefixed by the starting time. Instructions inside a rectangular box are executed sequentially. Additionally, two rectangular boxes can be executed concurrently if sufficient resources are available on grid, and as long as the instructions that they depend on are already completed. According to the input sequence graph of the given conditional assay shown in Fig. 7, two reagents R_1 and R_2 are mixed by the mix operation M_1. In the actuation sequence shown in Fig. 7, at time step 1, two droplets of reagents R_1 and R_2 are dispensed at locations $(1,8)$ and $(1,11)$ on the grid respectively. After dispensing, the droplets move to the mixer placed at location $(4,8,4,11)$. The instructions at time steps 2 to 4 are used to move the droplets from their dispense location to the mixer. At time step 5, the mixer starts to mix and the mixing continues for 4 time cycles. At time step 9, the mixer produces two output droplets at locations $(4,8)$ and $(4,11)$. There is a checkpoint placed at location $(4,8)$ which measures the concentration factor of the produced output droplet. After the mix operation, the output droplet stays at location $(4,8)$ and the other output droplet stays at location $(4,11)$. The output droplet at location $(4,11)$ starts to move at time step 10 and reaches the final output reservoir FO_1 placed at location $(6,12)$ at time step 13. Hence, the movement instructions $(m[(4,11) \longrightarrow (5,11)])$ from time step 10 to 13 are used to move one droplet from its source location to the output reservoir FO_1. Concurrently, on the other side, two droplets of reagents

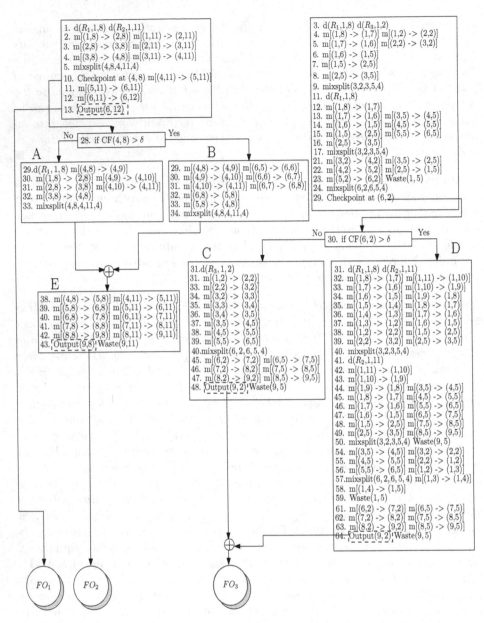

Fig. 7. Actuation sequence of the conditional assay (Color figure online)

R_1 and R_3 are dispensed at locations $(1,8)$ and $(1,2)$ respectively at time step 3. These two droplets move to the mixer at location $(3,2,3,5)$. These two droplets start to move at time step 4. The droplet of reagent R_1 starts to move from its source location $(1,8)$ at time step 4 and reaches the mixer input location $(3,5)$ at time step 8. The droplet of reagent R_3 starts to move from its location $(1,2)$ at *time step 4 and reaches the mixer input location* $(3,5)$ *at time step 5. The mixer at location* $(3,2,3,5)$ *starts to mix at time step 9 and mixes for 4 time cycles. The mixer finishes its mix operation at time step 12. Two output droplets are produced after the mix operation at locations* $(3,2)$ *and* $(3,5)$*. The output droplet at location* $(3,5)$ *moves to one input pin* $(6,5)$ *of another mixer placed at location* $(6,2,6,5)$*. At time step 11, a droplet of reagent* R_1 *is dispensed at location* $(1,8)$ *and moves to the mixer at* $(3,2,3,5)$*. The droplet at location* $(1,8)$ *reaches the input pin* $(3,5)$ *of the mixer placed at location* $(3,2,3,5)$ *at time step 16. At time step 17, the mixer* $(3,2,3,5)$ *starts to mix and mixes for 4 time cycles. The mixer produces two output droplets at locations* $(3,2)$ *and* $(3,5)$ *at time step 20. The output droplet at location* $(3,2)$ *moves to the input pin* $(6,2)$ *of the mixer* $(6,2,6,5)$*. The other output droplet (/ waste droplet) at location* $(6,5)$ *moves to the waste reservoir* $(1,5)$ *at time step 23. The mixer at location* $(6,2,6,5)$ *starts to mix at time step 24 and mixes for 4 time cycles. The mixer produces two output droplets at locations* $(6,2)$ *and* $(6,5)$ *respectively at time step 28. There is a checkpoint at location* $(6,2)$ *that measures the concentration factor of the output droplet on that location. Till now the flow of execution of the set of instructions is same as the straight line execution of the classical assay. But now there is something new in case of the conditional assay, where the measured concentration factor of the output droplets at certain locations by the checkpoints are checked against certain conditions. At time step 28, the measured concentration factor of the droplet at location* $(4,8)$ *is compared against some predefined value* δ*. If the concentration factor of the droplet at location* $(4,8)$ *is greater than* δ*, the set of instructions in the rectangular box B is executed, else the set of instructions within the rectangular box A is executed. This is symbolized notionally with a* $\oplus$ *node in the figure to indicate that the input received to the subsequent box E is generated either from block A or from block B. If the condition is satisfied, the output droplet at location* $(6,5)$ *of the mixer* $(6,2,6,5)$ *moves to the mixer at* $(4,8,4,11)$*. The mixer at* $(4,8,4,11)$ *starts to mix at time step 34 and produces its outputs at time step 37. If the condition at the checkpoint at* $(4,8)$ *is not satisfied, a droplet of the reagent* R_1 *is dispensed at location* $(1,8)$ *at time step 29 and it moves to the mixer at location* $(4,8,4,11)$*. The mixer starts to mix at time step 33 and produces output droplets at time step 36. Now, the output droplet at location* $(4,8)$ *starts to move at time step 38 and reaches the output reservoir* FO_2 *at location* $(9,8)$ *at time step 43. Another output droplet at location* $(4,11)$ *starts to move at time step 38 and reaches the waste reservoir at* $(9,11)$ *at time step 43. On the other hand, at time step 30, the measured concentration factor of the output droplet at location* $(6,2)$ *by the checkpoint is compared with the value* δ*. If the concentration factor is greater than* δ*, the set of instructions within block D is executed, otherwise the set of instructions within box C is executed. Hence, the set of instructions within box C or D is executed and the final output droplets are stored at the output reservoirs at* $(9,2)$ *and* $(3,1)$ *respectively. If the evaluation of the condition at time step 30 is true, then the execution of the assay completes at time step 64 and we get an output droplet at the final output reservoir* FO_3 *placed at location* $(9,2)$*. Otherwise, the execution of the assay completes at* *time step* 48 *and produces only one output at the final reservoir* (9, 2). *This is symbolized notionally with a* ⊕ *node in the figure, at the input of* FO_3. *At the end of execution of the assay, we get 3 final output droplets stored at the output reservoirs* FO_1, FO_2 *and* FO_3, *which are placed at locations* (6, 12), (9, 8) *and* (9, 2) *respectively. Out of the final output droplets, the droplet at reservoir* FO_3 *at location* (9, 2) *is assumed to be erroneous. We aim to debug the actuation sequence to identify the root cause of the erroneous output.* ■

The above example demonstrates the execution of a conditional assay, where the instructions to be executed are based on some decisions taken at run-time based on the outcome of checkpoints. Errors in such conditional assays are more difficult to debug than their counterparts. In the following discussion, we present our methods for debugging such errors.

4.1 Debug Methodology

As in the case of debugging unconditional assays, the debug methodology for conditional assays also consists of three steps, namely (a) Time frame expansion, (b) Slice Computation, and (c) Error localization. The methods have a similar spirit as earlier, however, with certain modifications that are needed to handle the conditional assay nodes. We explain each step in the discussion below.

Time Frame Expansion of Conditional Assays: Figure 6 shows a conditional assay description. As stated earlier, the time frame expansion step expands the bio-assay over time on the input grid. Here, the given conditional bio-assay is realized on a grid of size 9 × 12. Figure 8 shows the time frame expanded graph of the conditional assay. While the semantics of most of the nodes are same as earlier, we elaborate on the points of difference through the following illustration. The elliptical nodes annotated with a "check" condition, and the circular nodes annotated with ⊕ are different from the usual nodes present in case of unconditional assays, as shown in Fig. 3. As earlier, the time expanded graph of the conditional assay also consists of some dispense nodes (D_{id}^t, where id shows the droplet id and t represents the time when the droplet is dispensed), move nodes (mv_{id}, where id represents the id of the movement operation) and mix nodes (M_{id}, where id represents the id of the mix operation). These nodes are annotated with the start and end times of the corresponding operations. The new nodes in conditional assays, are the *check* nodes *check* O_{id} and the xor nodes ⊕. Here, $checkO_{id}$ represents the checkpoint operation, that evaluates a certain condition (e.g. concentration of O_{id} is less than a value) on the output O_{id} (where, id represents the output id) as observed at the checkpoint. These check nodes are also annotated with the time t_{check} that shows the time when the output is observed and checked at that checkpoint. Since conditional assays consist of branching on the assay execution depending on the evaluation of certain conditions, the direction of assay execution when a check node is encountered, is decided at run-time. For paths which depict such an operation where one among

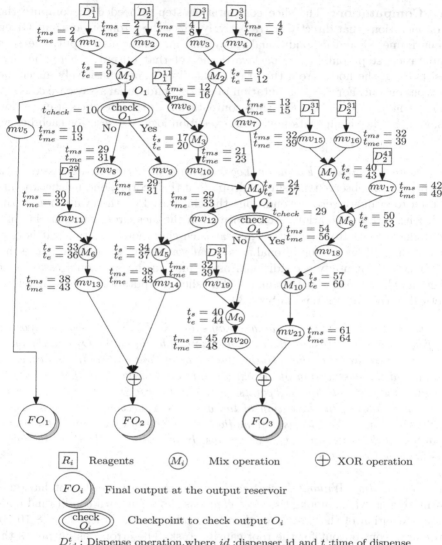

Fig. 8. Unfolded Input sequence graph of a conditional bio-assay

two outputs needs to be depicted, we add a xor operation denoted by $\oplus$ in the time expanded graph to depict that only one among the block of instructions between the *if* or *else* block is executed to produce the output.

Slice Computation: The slice computation step is needed to compute the set of operations that directly or transitively affect the erroneous output. Given a time frame expanded conditional assay and an erroneous output, the slice computation step computes a backward slice of that output, where the slice consists of all the nodes from the graph that directly or transitively affect the erroneous output. For slice computation in case of conditional assays, we consider the following cases, namely (A) slice computation in the presence of execution logs, and (B) slice computation without execution logs. We describe each below.

Slice Computation with Execution Logs: In case of a conditional assay with checkpoints, the observed measure of droplets at their corresponding checkpoints are used to evaluate certain conditions therein. Based on the evaluation of the conditions, certain operations are performed. If the measures at the checkpoints are available to us from execution logs at such checkpoint locations, it is easy to trace the path of the conditional assay that was taken at run-time. Once this is done, we can simplify a conditional assay into an unconditional assay since either the *if* or the *else* path remains. The slice computation step is therefore, identical to the one outlined in Sect. 3.

Example 4. Consider the conditional assay shown in Fig. 8 with two checkpoints. The first checkpoint measures the concentration factor of output O_1 and the other checkpoint measures the concentration factor of the output O_4. In this case, we assume that the observation of the checkpoints are known to us. Let us assume both the conditions at the checkpoints evaluated to true in a certain execution. Once this is known, the execution of the assay is reduced to the one shown in Fig. 9. Again, as earlier, let us assume that at the end of execution of the assay, the output at the reservoir FO_3 is erroneous. In this case, the debug methodology is as described in Sect. 3. ∎

Slice Computation Without Execution Logs: In this case, we do not have any information on which among the *if* or *else* was chosen at each conditional node during execution of the assay. We therefore, compute the static slice [8,10,20] of the erroneous output for the root-causing task. The erroneous output is the slicing criteria and slice computation is performed on the time frame expanded graph of the conditional assay. The algorithm is similar in spirit to the one described in Sect. 2. It takes an output node as the input (c) and works on the time expanded graph of the conditional assay and produces an annotated slice as the output that also keeps track of the guarding check conditions as described below. At the initialization of the algorithm, the output variable slice is initialized to empty (ϕ). Starting from an output (slicing criteria), this algorithm performs a backward traversal on the graph to compute the slice, as earlier. We traverse the graph and check all operations of the assay that are executed before the slicing criteria c. Any operation OP that influences the slicing criteria directly or transitively is included in the slice. The operations that may be encountered at the time of traversing the graph are *dispense, move, mix, check* and $\oplus$.

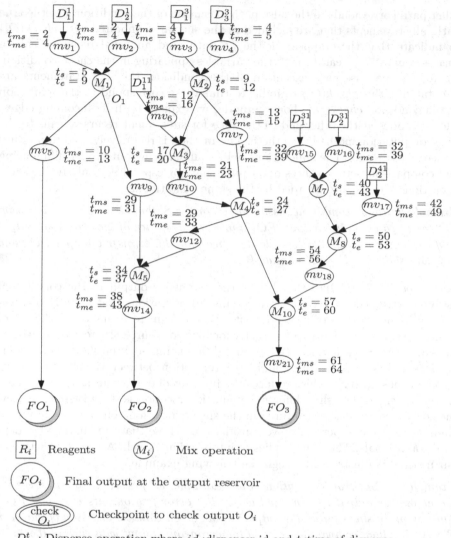

Fig. 9. Time expanded graph with the execution log of the checkpoint observation

Except the operations *check* and $\oplus$, slice computation for the other operations *dispense, move* and *mix* are exactly identical to the sliceComputation(c) method described earlier in the previous section. The operation $check(O_i)$ used for a condition checking operation, needs special attention. Here, there are two paths of execution. One path is for the true path execution of the condition and the other path corresponds to the false path execution of the condition. For the true path, all statements that are included in the slice are annotated with $check(O_i)$ to indicate that these appear in the if path, and an usual recursive call in turn is generated as earlier with the variables appearing in the check condition. Again, for the false path execution of the condition, the slice statements are annotated with $\neg check(O_i)$ to indicate that these correspond to the refutation of the condition $check(O_i)$. For a $\oplus$ node, we explore both its incoming edges, add the nodes at the other end to the slice for subsequent recursive calls to the slice computation method with these. The method is recursively used to produce a slice as the result. The annotations help us in carrying out proper differencing and comparison between slices of outputs that correspond not only to the same operations but are also guarded by the same conditionals.

Example 5. In the conditional assay shown in Fig. 8, there is an erroneous output FO_3 and two error-free outputs FO_1 and FO_2. The slice of the erroneous output FO_3 is shown in Fig. 10(C), while the slices for the error-free output FO_1 and FO_2 are shown in Fig. 10(A) and (B). ∎

Error Localization: Once the slice of the erroneous output of the conditional assay is extracted, we have a set involving all operations that directly or transitively influence the erroneous output. We now aim to prune down the set of operations in the slice further to exactly localize the source of error. This involves the same steps as earlier, namely (a) Slice differencing between slices of erroneous outputs and error-free output, and (b) intersection between the slices of multiple erroneous outputs. This produces the final set of operations as the suspected sources of error. The slice differencing and intersection is done keeping in view the conditional nodes, as captured in the slice expression. Thus, for example, to eliminate common operations, we consider only those that are guarded by identical conditionals. The same applies to intersection as well. A visual depiction of our method is illustrated through the following example.

Example 6. For error localization in our example, we first compute the slices for the erroneous output FO_3 as well as for the error-free outputs FO_1 and FO_2. The computed slices for all the outputs of the given conditional assay is shown in Fig. 10. We then compare the slice of the erroneous output with each of the slices of the error-free outputs. We also have to update the slice of the erroneous output by differencing the common operations present in both the erroneous and error-free outputs. From Fig. 10, it is clear that there is no operation that is present in both the slices of the erroneous output FO_3 and the slice of the error-free output FO_1. However, in case of the comparison between the slices of the erroneous output FO_3 and the error-free output FO_2, we find a set of common operations. Hence, we remove the common operations from the slice of the erroneous output FO_3. The updated slice of the erroneous output FO_3 is shown in Fig. 11. In this example, there is only one erroneous output. Hence we are not able to reduce the error suspect region any further. ∎

In the following section, we discuss details of the implementation of our debug methodology along with our experiments.

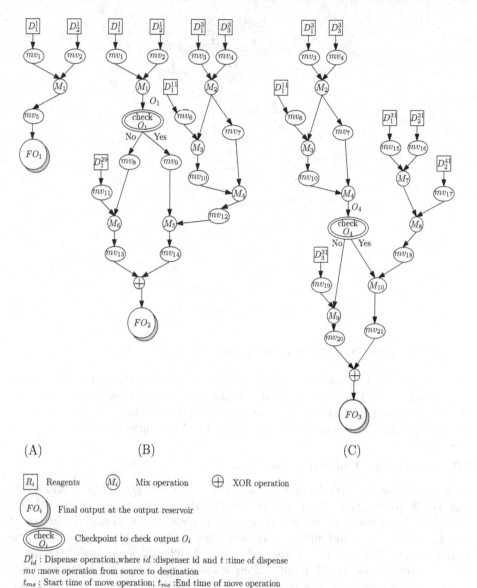

Fig. 10. (A) Slice for error-free output FO_1, (B) Slice for error-free output FO_2, (C) Slice for erroneous output FO_3

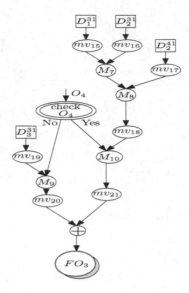

Fig. 11. Slice comparison

5   Implementation and Results

Our implementation takes as input a golden assay, the actuation sequence and the output error. It produces a set of operations that are potential error sources. We report experimental results after executing the method on two types of assays, namely, (A) Unconditional assays and (B) Conditional assays.

5.1   Results on Unconditional Assays

Experiments were carried out on the Linear Dilution Tree (LDT) and Polymerase Chain Reaction (PCR) assays. We first discuss our experiments on LDT. Figure 12 shows the LDT input graph. We considered a 8×13 grid as the target biochip architecture and an actuation sequence consisting of 19 operations produced by a synthesis tool. We injected errors at different locations of the assay, one at a time (single attack point and error source). We describe below one such experiment, where error was injected at mixer M_9. The effect of the erroneous mixer is seen at outputs O_5 and O_7 which are erroneous and shown as red circles in Fig. 12.

We carried out our steps of slice computation followed by differencing and intersection. Figure 13(i) and (ii) depicts the results of the different steps. Figure 13(i)(F) shows the final slice, which points towards an error in M_9. Our method is able to cut down the size of the suspect erroneous region significantly in this case, finally reducing to a single node. This shows its effectiveness. We randomly varied the attack location to test the application of our method. Column 3 shows the number of nodes present in the original assay, while Column 5 shows the number of nodes present in the final slice.

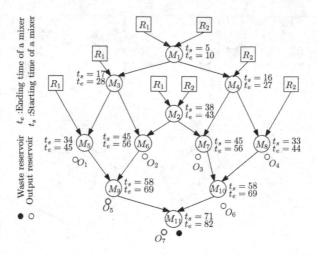

Fig. 12. Input sequence graph for LDT (Color figure online)

Table 1. Experimental results on LDT and PCR

Type of assay	Erroneous output	Nodes present on assay	Slice contains	After pruning remaining nodes
LDT	O_4	19	19	1
	O_1	19	7	4
	O_4		19	
	O_6		12	
	O_2	19	12	3
	O_3		9	
	O_4		12	
	O_2	19	12	8
	O_3		9	
	O_4		12	
	O_5		9	
	O_6		12	
PCR	O_1	28	15	2
Streaming	O_3	28	15	2
	O_1	28	15	1
	O_2		15	
	O_5	28	15	5

A similar error injection and debug experiment (as shown in Table 1) was carried out for the PCR Streaming protocol. In this case as well, our method could reduce the suspect region considerably.

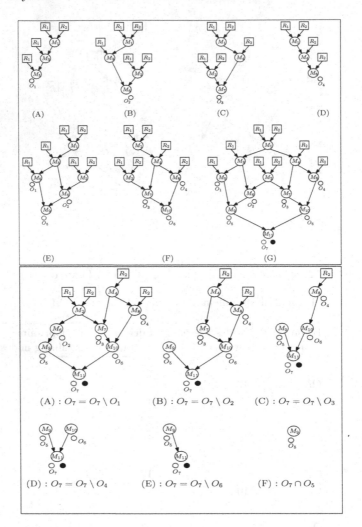

Fig. 13. (i) Slice computation (ii) Error Localization steps

5.2 Debugging Conditional Assays

We first report the results of our experiments on some synthetic conditional assays followed by a variant of the LDT assay with conditional nodes depicting an *if-else* structure denoting re-synthesis and relocations of mix operations based on checkpoint observations.

Synthetic Conditional Assays: Figure 14 shows a synthetic conditional assay, that has been constructed with suitable modifications to a classical solubility reaction. In the assay, two reagents R_1 and R_2 are mixed by a mix operation M_1. One output of the mix operation is mixed with the reagent R_3 by the mix

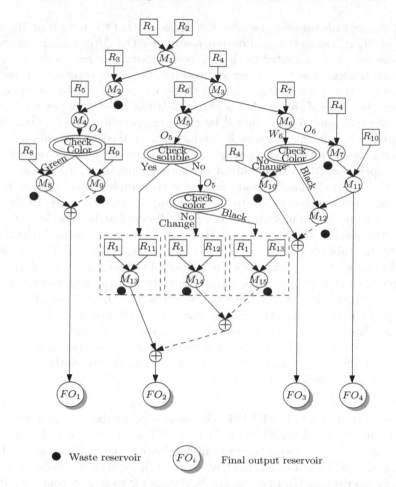

Fig. 14. Input Sequence Graph of Conditional synthetic assay

operation M_2 and the other output is mixed with reagent R_4 by the mix operation M_3. One output of the mix operation M_2 is mixed with the reagent R_5 by the mix operation M_4. The other output droplet of the mix operation M_2 moves to the waste reservoir. In the same way, one output of the mix operation M_3 is mixed with the reagent R_6 by the mix operation M_5 and the other output droplet is mixed with the reagent R_7 by the mix operation M_6. The outputs of the mix operations M_4, M_5 and M_6 are checked by their respective checkpoints. The checkpoints observe certain parameters of the output droplet like, color, solubility, concentration factor of the droplet etc. Then the execution path is decided based on the observation of the checkpoints, as is the case for a conditional assay. After the mix operation M_4, the color of the residue is checked. If the color is green, then the reagent R_8 is mixed with the output of M_4 by the mix operation M_8, else the output of M_4 is mixed with the reagent R_9 by the mix operation M_9. The evaluation of the condition checked at the checkpoint decides the next operation to be executed. Hence, either the output of M_8 or the output of M_9 moves to the final output reservoir FO_1. After the mix operation M_5, a checkpoint is evaluated to check if there occurs any residue (not soluble) or soluble. If there does not occur any residue, then two reagents R_1 and R_{11} are mixed by the mix operation M_{13}. If there occurs any residue at the end of M_5, then the color of the residue is checked. If the color does not change, the two reagents R_1 and R_{12} are mixed by the mix operation M_{14}. If the color of the residue is black, the reagents R_1 and R_{13} are mixed by the mix operation M_{15}. Now, based on the checkpoint evaluation at run-time, either the output of the mix operation M_{13} or the output of the mix operation M_{14} or the output of the mix M_{15} is the final output and one of the droplets produced at the corresponding mixer output moves to the final output reservoir FO_2. In the same way, the residue of the mix operation M_7 is observed at the checkpoint. If the color of the residue does not change, the reagent R_4 is mixed with the residue of M_6 by the mix operation M_{10}. If the color of the residue of M_6 is observed as black, the residue is mixed with the output of the mix operation M_{11} by the mix operation M_{12}. On the other hand, the filtrate of the mix operation M_6 is mixed with the reagent R_4 by the mix operation M_7. One output of M_7 moves to the waste reservoir and the other output is mixed with the reagent R_{10} by the mix operation M_{11}. One output of M_{11} moves to the final output reservoir FO_4 and the other output of M_{11} is mixed by the mix operation M_{12}. In this case as well, either the output of M_{10} or the output of M_{12} is the final output and moves to the final output reservoir FO_3.

A Conditional Variant of LDT: We now explain the structure of a modified variant of the Linear Dilution Tree (LDT) assay (shown in Fig. 12) with conditional nodes, which do not exist in its classical description. To create the conditional assay for our experiment, we built on the idea of relocation and resynthesis, as described in [7]. In other words, the idea is to have certain conditional statements in the assay that have an *if* branch that includes the statements in the classical description and an *else* part that includes statements to relocate and resynthesize at execution time in case of a physical error or an operational ambiguity. As an example, let us suppose that the physical location of a mixer on the grid is erroneous, hence, the mixer location is re-configured and then the relevant portion of the assay is re-synthesized. Figure 15 shows a LDT assay description annotated with certain conditionals. In the assay, two reagents R_1 and R_2 are mixed by the mix operation M_1. The output droplets O_1 and O_2 of M_1 are checked against certain conditions to judge the quality of the mix. If there is any error, then the mixer is reconfigured at another location on the grid and that part of the assay is re-synthesized. The operations that correspond to the resynthesis are enclosed inside a dashed rectangular box. After resynthesis, the mix operation M_1 is represented by the mix operation M_1'. Output droplets O_1' and O_2' are produced after the mix operation M_1'. If the outputs O_1 and O_2 of M_1 are error-free, as evaluated with a check, the mixer reconfiguration is not needed. Hence, either the output O_1 or O_1' is mixed with the reagent

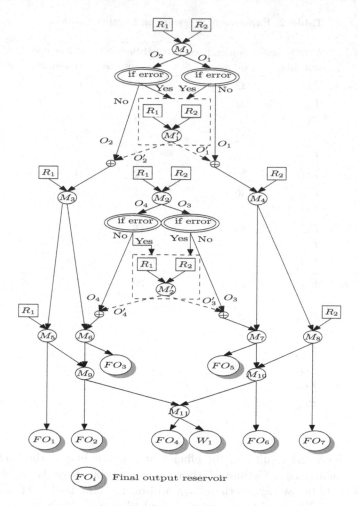

Fig. 15. Input Sequence Graph for Conditional LDT

R_2 by the mix operation M_4. In the same way, either the output O_2 or O_2' is mixed with the reagent R_1 by the mix operation. Again, there is a checkpoint at the output of the mix operation M_2. M_2 mixes two reagents R_1 and R_2. If the mixer is faulty, it implies that the outputs O_3 and O_4 of M_2 are erroneous and this leads to the mixer being relocated and the corresponding operations being re-synthesized. Now, if the operations are re-synthesized, the outputs O_3' and O_4' are produced by M_2'. Thus, in this case, only one among O_3 and O_3' is mixed with the output of M_3. Similarly, one among O_4 and O_4' is mixed with the output of the mix operation M_7. At the end of the execution of the assay, the final outputs are stored at the final output reservoirs FO_1, FO_2, FO_3, FO_4, FO_5, FO_6 and FO_7 respectively. For our experiments, we randomly varied the erroneous output location and then recorded how our method could cut down the size of the slice. Experimental results show the efficiency of our method.

Table 2. Experimental results on Synthetic assays

Type of assay	Erroneous output	Nodes present on assay	Slice contains	After pruning remaining nodes
Synthetic 1	FO_3	22	15	9
	FO_1	22	3	3
	FO_2		13	
	FO_2	22	13	6
	FO_3		15	
Synthetic 2	FO_1	36	12	9
	FO_3	36	15	2
	FO_4		11	
	FO_2	36	18	2
	FO_3		15	
	FO_4		11	
Conditional LDT	FO_4	29	29	1
	FO_1	29	12	2
	FO_2		22	
	FO_4		29	
	FO_4	29	29	1
	FO_5		19	
	FO_6		22	

Table 2 shows the results on the efficiency of our debug method for debugging conditional assays. Column I shows the type of assay. Here, we report our experiments on two synthetic assays and on the standard LDT assay with the checkpoints. The conditional assay described in Fig. 6 in Sect. 4 is referred as $Synthetic1$ in Table 2. The assays $Synthetic2$ and $ConditionalLDT$ are as described earlier in this section. Table 2 shows the performance of our method in terms of the size (in terms of the number of operations) of the suspected error region produced as the final result in the final column.

Performance Study of Our Debug Method: Further, we varied the size of the grid for all the assays and the error source location simultaneously to study the scalability of our method. We injected errors at different locations of the assay and analyzed the performance of our tool. In each case, the size of the suspected region was significantly less as compared to the original assay. Performance experiments with our tool are discussed in Table 3. The first column of the table shows the type of the assay. Column 2 and Column 3 show the index of the experiments of each protocol and the grid sizes of each protocol respectively. Column 4 shows the injected error location for each experiment and

Table 3. Performance records on LDT and PCR with varying grid sizes

Type of Assay	Index	Grid size	Injected Error	Erroneous Output	Time for pruning (sec)	Peak memory (MB)
LDT	1	8×13	M_{11}	$(8, 7)$	61	0.177
			M_8	$(8, 1)$ $(8, 7)$ $(8, 11)$	120.68	0.177
			M_6	$(8, 3)$ $(8, 5)$ $(8, 7)$	360.85	0.177
			M_2	$(8, 3)$ $(8, 5)$ $(8, 7)$ $(8, 9)$ $(8, 11)$	48	0.158
	2	10×13	M_{11}	$(10, 8)$	360	0.185
			M_8	$(10, 2)$ $(10, 8)$ $(10, 12)$	1583	0.215
			M_5	$(8, 13)$ $(10, 4)$ $(10, 8)$	1380.96	0.216
PCR Streaming	1	8×15	M_{13}	$(8, 3)$ $(8, 3)$	281	2.55
			M_{12}	$(8, 8)$ $(8, 8)$	281	2.55
			M_{14}	$(6, 3)$ $(6, 3)$	1680.09	4.1
			M_{11}	$(8, 13)$ $(8, 13)$	1184	4.1
	2	10×17	M_{12}	$(10, 8)$ $(10, 8)$	477	3.1
			M_{14}	$(6, 3)$ $(6, 3)$	1689	4.1
			M_{11}	$(10, 13)$ $(10, 13)$	645	2.55
Synthetic1	1	10×12	M_1	$(6, 12)$ $(10, 8)$	576	0.216
			M_8	$(10, 2)$	382	0.172
	2	8×12	M_1	$(6, 12)$ $(8, 8)$	278	0.177
			M_7	$(8, 2)$	221	0.158
Synthetic2	1	10×16	M_4	$(10, 3)$	448	0.184
			M_6	$(10, 9)$ $(10, 14)$	1120.36	0.241
	2	12×18	M_4	$(12, 3)$	676.12	1.2
			M_3	$(12, 6)$ $(12, 9)$ $(12, 16)$	1765	4.1
Conditional LDT	1	12×20	M_{11}	$(12, 11)$	1420	3.1
			M_9	$(12, 5)$ $(12, 11)$	1680	4.1
	2	14×20	M_8	$(14, 11)$ $(14, 17)$ $(14, 20)$	1826.2	4.23

Column 5 shows the erroneous output locations. Finally, the last two columns of the table respectively show the time needed for the pruning method and the peak memory required for the respective experiments.

We now compare our results with the one in [19]. In this work, the authors have proposed an error recovery technique that re-synthesizes the suspected error region. For finding the suspected error region starting from an error manifestation, the slice of the error is computed. In Table 1, we have shown the number of statements that remain as suspected error origin locations, after the slice is computed. Our debugging method not only computes the slice of the erroneous output but also performs a slice comparison step for pruning the nodes which cannot affect the erroneous output. Hence, our method provides lesser number of nodes in the final result as possible origins of the erroneous output, as evident from the numbers in the final column of the same table.

6   Conclusion

In this paper, we propose an error debugging method that can efficiently localize errors in DMF executions for both unconditional assays that have been heavily used in literature, and their conditional counterparts that include the flexibility of incorporating error recovery and resynthesis in the assay description. Our method is completely automatic and can localize the earliest error location in the erroneous assay using backward slicing and slice comparison. While the method proposed in this paper is carried out as an offline step, it can also be extended for online error localization. This would make our method more effective in practice since significant time and resources can be saved by doing a timely online error detection and re-synthesis.

Acknowledgement. This work was supported by a grant received from the Science and Engineering Research Board (SERB), Government of India, through an extramural research project EMR/2016/005977.

References

1. Ali, S.S., et al.: Security implications of cyberphysical digital microfluidic biochips. In: ICCD (2015)
2. Ali, S.S., et al.: Security assessment of cyberphysical digital microfluidic biochips. TCBB **13**(3), 445–458 (2016)
3. Ananthanarayanan, V., Thies, W.: Biocoder: a programming language for standardizing and automating biology protocols. J. Biol. Eng. **4**(1), 13 (2010)
4. Banerjee, A., Roychoudhury, A., Harlie, J.A., Liang, Z.: Golden implementation driven software debugging. In: Proceedings of the Eighteenth ACM SIGSOFT International Symposium on Foundations of Software Engineering, pp. 177–186. ACM (2010)
5. Bhattacharjee, S., Banerjee, A., Chakrabarty, K., Bhattacharya, B.B.: Correctness checking of bio-chemical protocol realizations on a digital microfluidic biochip. In: 2014 27th International Conference on VLSI Design and 2014 13th International Conference on Embedded Systems, pp. 504–509. IEEE (2014)

6. Bhattacharjee, S., Banerjee, A., Ho, T.Y., Chakrabarty, K., Bhattacharya, B.B.: On producing linear dilution gradient of a sample with a digital microfluidic biochip. In: 2013 International Symposium on Electronic System Design (ISED), pp. 77–81. IEEE (2013)
7. Chakrabarty, K., et al.: Digital Microfluidic Biochips - Synthesis, Testing, and Reconfiguration Techniques. CRC Press, Boca Raton (2007)
8. Chang, J., Richardson, D.J.: Static and dynamic specification slicing. In: Proceedings of the Fourth Irvine Software Symposium (1994)
9. Chen, Y.H., Hsu, C.L., Tsai, L.C., Huang, T.W., Ho, T.Y.: A reliability-oriented placement algorithm for reconfigurable digital microfluidic biochips using 3-D deferred decision making technique. IEEE Trans. Comput.-Aided Des. Integr. Circuits Syst. **32**(8), 1151–1162 (2013)
10. Choi, J.D., Ferrante, J.: Static slicing in the presence of goto statements. ACM Trans. Program. Lang. Syst. (TOPLAS) **16**(4), 1097–1113 (1994)
11. Grissom, D., Brisk, P.: Path scheduling on digital microfluidic biochips. In: 2012 49th ACM/EDAC/IEEE Design Automation Conference (DAC), pp. 26–35. IEEE (2012)
12. Grissom, D.T., Brisk, P.: Fast online synthesis of digital microfluidic biochips. IEEE Trans. Comput.-Aided Des. Integr. Circuits Syst. **33**(3), 356–369 (2014)
13. Ho, T.Y., Zeng, J., Chakrabarty, K.: Digital microfluidic biochips: a vision for functional diversity and more than Moore. In: Proceedings of the International Conference on Computer-Aided Design, pp. 578–585. IEEE Press (2010)
14. Hu, K., Bhattacharya, B.B., Chakrabarty, K.: Fault diagnosis for leakage and blockage defects in flow-based microfluidic biochips. IEEE Trans. Comput.-Aided Des. Integr. Circuits Syst. **35**(7), 1179–1191 (2016)
15. Hu, K., Yu, F., Ho, T.Y., Chakrabarty, K.: Testing of flow-based microfluidic biochips: fault modeling, test generation, and experimental demonstration. IEEE Trans. Comput.-Aided Des. Integr. Circuits Syst. **33**(10), 1463–1475 (2014)
16. Jaress, C., Brisk, P., Grissom, D.: Rapid online fault recovery for cyber-physical digital microfluidic biochips. In: 2015 IEEE 33rd VLSI Test Symposium (VTS), pp. 1–6. IEEE (2015)
17. Keszocze, O., Wille, R., Drechsler, R.: Exact routing for digital microfluidic biochips with temporary blockages. In: Proceedings of the 2014 IEEE/ACM International Conference on Computer-Aided Design, pp. 405–410. IEEE Press (2014)
18. Keszocze, O., Wille, R., Ho, T.Y., Drechsler, R.: Exact one-pass synthesis of digital microfluidic biochips. In: Proceedings of the 51st Annual Design Automation Conference, pp. 1–6. ACM (2014)
19. Luo, Y., Chakrabarty, K., Ho, T.Y.: Error recovery in cyberphysical digital microfluidic biochips. IEEE Trans. Comput.-Aided Des. Integr. Circuits Syst. **32**(1), 59–72 (2013)
20. Mao, X., Lei, Y., Dai, Z., Qi, Y., Wang, C.: Slice-based statistical fault localization. J. Syst. Softw. **89**, 51–62 (2014)
21. Mazutis, L., Gilbert, J., Ung, W.L., Weitz, D.A., Griffiths, A.D., Heyman, J.A.: Single-cell analysis and sorting using droplet-based microfluidics. Nat. Protoc. **8**(5), 870 (2013)
22. Roy, P., Banerjee, A.: A new approach for root-causing attacks on digital microfluidic devices. In: AsianHOST, pp. 1–6 (2016)
23. Roy, P., Banerjee, A.: Security assessment of synthesized actuation sequences for digital microfluidic biochips. In: 7th International Symposium on Embedded Computing and System Design. ISED 2017, India, pp. 1–4 (2017)

24. Roy, P., Banerjee, A., Bhattacharya, B.B.: Debugging errors in microfluidic executions. In: Chaki, R., Cortesi, A., Saeed, K., Chaki, N. (eds.) Advanced Computing and Systems for Security. AISC, vol. 996, pp. 143–158. Springer, Singapore (2020). https://doi.org/10.1007/978-981-13-8969-6_9

25. Roy, S., Kumar, S., Chakrabarti, P.P., Bhattacharya, B.B., Chakrabarty, K.: Demand-driven mixture preparation and droplet streaming using digital microfluidic biochips. In: Proceedings of the 51st Annual Design Automation Conference, pp. 1–6. ACM (2014)

26. Sista, R., et al.: Development of a digital microfluidic platform for point of care testing. Lab Chip 8(12), 2091–2104 (2008)

27. Su, F., Chakrabarty, K.: High-level synthesis of digital microfluidic biochips. ACM J. Emerg. Technol. Comput. Syst. (JETC) 3(4), 1 (2008)

28. Su, F., Hwang, W., Chakrabarty, K.: Droplet routing in the synthesis of digital microfluidic biochips. In: 2006 Proceedings of Design, Automation and Test in Europe. DATE 2006, vol. 1, pp. 1–6. IEEE (2006)

29. Thies, W., Urbanski, J.P., Thorsen, T., Amarasinghe, S.: Abstraction layers for scalable microfluidic biocomputing. Nat. Comput. 7(2), 255–275 (2008)

30. Wu, P.H., Bai, S.Y., Ho, T.Y.: A topology-based eco routing methodology for mask cost minimization. In: 2014 19th Asia and South Pacific Design Automation Conference (ASP-DAC), pp. 507–512. IEEE (2014)

31. Xu, T., Chakrabarty, K.: Integrated droplet routing in the synthesis of microfluidic biochips. In: Proceedings of the 44th Annual Design Automation Conference, pp. 948–953. ACM (2007)

32. Yeh, S.H., Chang, J.W., Huang, T.W., Yu, S.T., Ho, T.Y.: Voltage-aware chip-level design for reliability-driven pin-constrained EWOD chips. IEEE Trans. Comput.-Aided Des. Integr. Circuits Syst. 33, 1302–1315 (2014)

33. Zhao, Y., Xu, T., Chakrabarty, K.: Integrated control-path design and error recovery in the synthesis of digital microfluidic lab-on-chip. JETC 6(3), 11 (2010)

Author Index

Printed in the United States
By Bookmasters